Geology of the Mahanoy, Mahantongo and Lykens Valleys

Something About The Earth We Walk On

Steve E. Troutman

SUNBURY PRESS
Mechanicsburg, Pennsylvania USA

Published by Sunbury Press, Inc.
105 South Market Street
Mechanicsburg, Pennsylvania 17055

www.sunburypress.com

ISBN: 978-1-62006-519-8 (Trade Paperback)

SECOND SUNBURY PRESS EDITION: November 2015

Product of the United States of America
0 1 1 2 3 5 8 13 21 34 55

Set in Bookman Old Style
Designed by Lawrence Knorr
Cover by Lawrence Knorr
Cover Art by Charles Knight – ***Orohippus and Uintatherium*** **-** Small four-toed horses of the genus *Orohippus* and the much larger, but harmless *Uintatherium* of the Eocene. ID # CK46T Charles R. Knight paintings from the Photo Archives. All images shown in the gallery are © The Field Museum. Photograph by Ron Testa or John Weinstein. *This idyllic scene painted by Charles R. Knight depicts central Pennsylvania during the early Tertiary Period. The Tertiary Period began 67 million years ago. The oldest part of this period is named the Eocene Epoch. The Eocene Epoch is pictured here. The climate was warm and moist. The dinosaurs of an earlier time had disappeared. Mammals of many different species became more prevalent. Here we see the small 4-toed horses Orohippus playfully scampering about. The much larger, but harmless, Uintatherium are grazing contentedly. During the Tertiary Period, the ridges and valleys of our central Pennsylvania landscape were shaped to form the Earth's surface we see today and walk upon.*
Edited by Lawrence Knorr

Continue the Enlightenment!

Dedication

This book is dedicated to the memory of Peter Craig Breen, son of Kevin and Renee Romberger Breen. His father, Kevin, graciously consented to assist with the draft manuscript. Kevin agreed to the dedication and provided these photos and captions.

The caption relating to the Romberger farm photograph (below) ties into the subtitle theme of the book ... *Something about the Earth We Walk on.*

Peter had a captivating smile (photograph on right taken December 2013) and was a joy to everyone who knew and worked with him. He earned his Bachelor of Science degree in Chemistry with Honors in 2014. Peter died on March 11, 2014, at the young age of 22.

The boys in the top photograph are walking in a northerly direction close to the Tulpehocken Path on remnants of the Ice Age terrace near Klingerstown where the Mahantongo, Pine, and Snow Creeks join. They are exploring the fields, rocks, and streams of the Mahantongo Valley at the Roland Romberger farm in 2003. Roland's grandson, Peter Craig Breen, age 12, is on the right. Roland's grandson, Scott Patrick Breen, age 9, is on the left. By exploring the streams and climbing the mountains near the Romberger farm, Peter and Scott found a love of the outdoors and both developed an interest in science. With an inquisitive manner and a keen mind for chemistry, math, and music, Peter enrolled in the Chemistry Program at Pennsylvania State University in 2010 and served as teaching assistant and tutor in chemistry during his studies.

Thank You

Thank you to Franklin and Marshall College, Lancaster, PA, for my geology education. I graduated in 1974 with a Bachelor of Arts Degree, majoring in Geology. This book is a result of my interest in local geology.

I wish to give a special thank you to Kevin Breen of 507 Ellen Road, Camp Hill, PA 17011. Kevin is a geologist with the U.S. Geological Survey. Kevin and I became friends when he married my first cousin, Renee Romberger. Renee and I grew up as neighbors near Klingerstown.

In December of 2013, I saw Kevin and Renee at a Romberger family gathering at the Roland Romberger homestead, one half mile west of Klingerstown. I asked Kevin to look over a rough draft of this book and to comment on the content. In May of 2014 he replied, "Overall, I enjoyed reading the text and examining the illustrations and photographs. The content is suitable for an audience interested in local geology and history. The focus of my review was to add or update references to support the presentation." His suggestions to improve the draft manuscript and make the book more reader-friendly were important improvements.

I wish to give a special thank you to Lawrence K. Knorr of Sunbury Press. Sunbury Press is a publishing house based in Mechanicsburg, PA with Pennsylvania topics as one of its specialties. In December of 2013, Larry reviewed my manuscript. His initial response was encouraging. He replied, "I definitely want to do the geology book." This was the encouragement that I needed to complete the book.

Thank you to the Field Museum of Chicago, Illinois. I was graciously granted permission to use images from the Charles Knight murals in their collection. Mr. Knight created artwork depicting animal and plant life from prehistoric eras including dinosaurs, birds, and mammals. Although this art was created over fifty years ago, Mr. Knight's images are timeless to a geologist. His scenes depicting our world during the Paleozoic, Mesozoic, and Cenozoic Eras are presented in this book courtesy of the Field Museum.

Finally, thank you to my family for supporting me in this book publication. My dear wife, Joan E. Troutman typed most of this book. This publication would not exist without her contribution. In addition, my daughter, Valerie E. Specht, spent many hours providing the technical support needed to prepare the manuscript pages in digital format.

Geology of the Mahanoy, Mahantongo, and Lykens Valleys
Something About the Earth We Walk On
by Steve E. Troutman

Geology is the study of the Earth's surface and the rock types and structures that form this surface. This is the surface of the land that we see every day and walk upon. As an introduction I would like to consider the oldest written record of the creation of the Earth. It is found in the Bible, Genesis, Chapter 1, verses 1-27:

In the beginning God created the heaven and the earth.

On the first day the Earth was without form. And God said, "Let there be light."

On the second day God made the firmament and divided the waters which were under the firmament, (Earth), from the waters that were above the firmament, (Heavens).

On the third day God said, "Let the waters under the heaven be gathered together in one place, and let dry land appear." And God said, "Let the earth bring forth grass and herbs".

On the fourth day God said, "Let the waters bring forth abundantly the moving creature that has life." He created the great whales and every living creature that moves within the waters.

On the fifth day God said, "Let the Earth bring forth living creatures after his kind, the cattle and creeping thing and beasts of the earth."

On the sixth day God said, "Let us make man in our image, after our likeness and let them have dominion over all the Earth."

On the seventh day, God finished His work which He had done in Creation, and He rested.

If we consider the Bible creation story and compare the Bible record to what modern day science and the study of geology proposes as the formation of the Earth, we will find that the creation story is essentially the same. The only difference is that geology uses a different time scale as a method of accounting for this time during the creation of the earth. The seven days described in the Bible equate to 4 billion years on the geologic time scale describing the age of the Earth.

The story of creation described in the Bible begins with formation of light out of darkness, followed by the formation of the oceans, dry land, plant life, animal life and last of all, human life. This is the same story supported by our study of Earth Science.

Geology

All who have studied the Earth as geologists, have concluded that a basic principle exists which is named, Uniformitarianism. This is a big word which simply means that the Earth shaping processes which we observe today have also been in place during the Earth's past. Earthquakes, wind, rain, freeze and thawing, moving ice, and flowing water, all are presently active forces changing the face of the Earth today, and have also done so in the past.

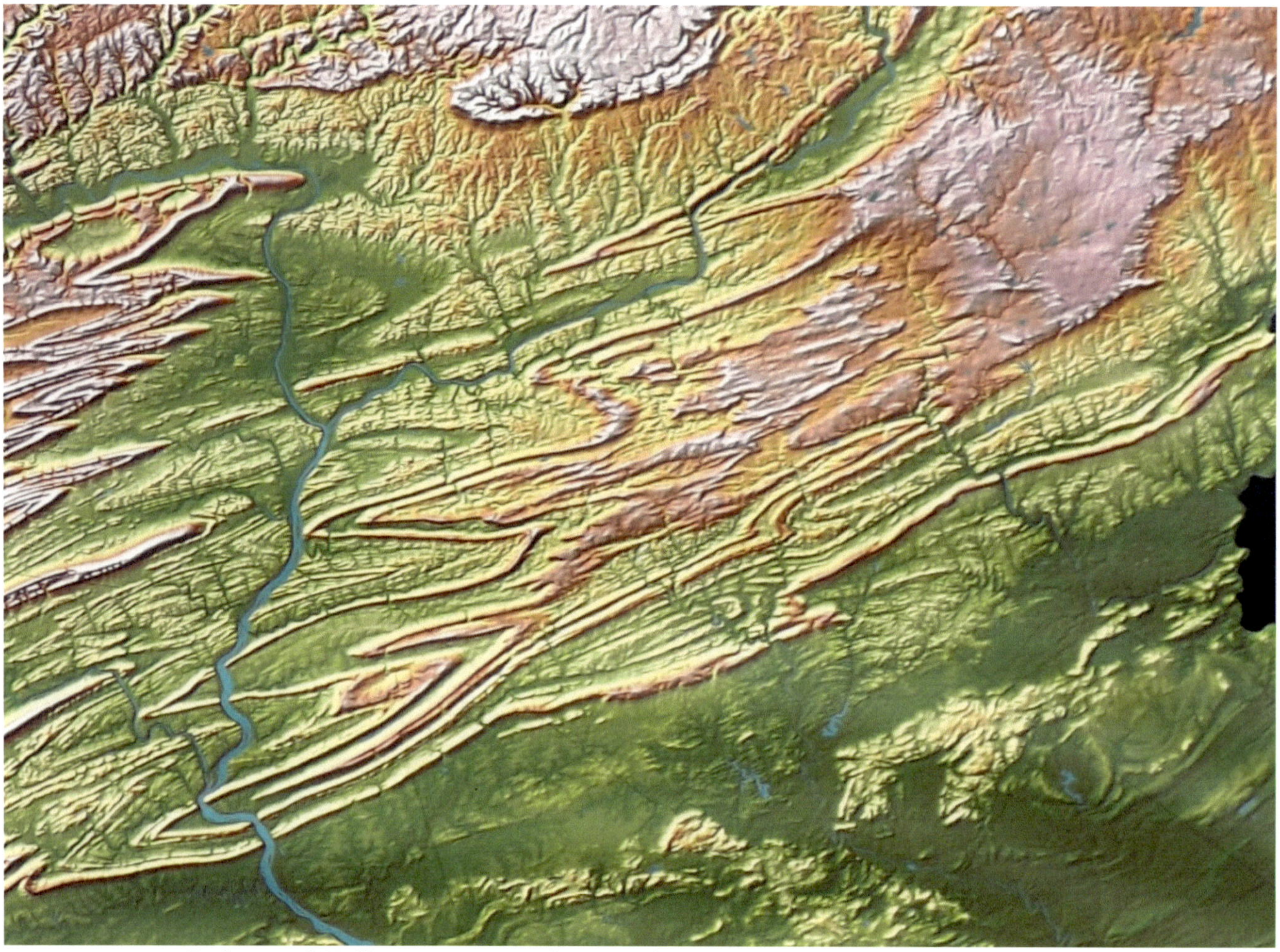

Ridge and Valley Province of Pennsylvania, Pa. Geological Survey Map 65http://www.dcnr.state.pa.is/cs/groups/public/documents/document/dcnr_015423.zip

Let us now consider our local area comprised of the Mahanoy, Mahantongo and Lykens Valleys (see photo on the next page). Within the larger Mahantongo Valley is contained the Hoofland Valley. Within in the Lykens Valley is contained the coal bearing rock layers of Bear Valley which is north of Williamstown. Within the Mahanoy Valley is contained the coal bearing rock near Trevorton.

All the rocks in the area of our study are sedimentary. These rocks all originated within water or were associated with water as their environment of deposition. Red mud like is found on the Mississippi Delta, forms into red shale layers. Deep water sediments, such as in the Caribbean Sea today, are calcium rich and become limestone. Sands like those found at the ocean beaches are turned into sandstone. These layers of sedimentary rocks are termed rock strata. Sandstone which contains larger sedimentary grains of various sizes including larger pebbles is named conglomerate.

The Appalachian Mountain Ridges of our area of study are formed by three major rock strata exposed as ten separate ridges.1) Little Mountain, Line Mountain, Mahantongo Mountain and Berry Mountain are formed from Pocono Sandstone Strata 2) Big Mountain and Mahanoy Mountain form the Treverton Coal Fields, Bear or Thick Mountain, Big Lick Mountain and Short Mountains forms Bear Valley Coal Fields which are formed of Pottsville Sandstone Strata.3) Hoofland Mountain and Fishers Ridge are formed from Montebello Sandstone Strata

As we look at the rock layers within our area of study, the oldest are found in the center of the Hoofland Valley. The youngest rocks are found in Bear Valley and Mahanoy Valley. We have a continuing progression from oldest to youngest sedimentary strata. If we were to walk in a southerly direction from the center of Hoofland Valley to the center of Bear Valley we would be crossing layers of bedrock which are increasingly younger and younger in age.

Similarly, if we were to walk in a northerly direction from the center of Hoofland Valley to the center of the Mahanoy Valley near Hunter Station, we would be crossing layers of bedrock which are increasingly younger and younger in age.

The limestones in the center of Hoofland Valley contain the earliest marine fossils such as tiny shells and crustaceans. Meckley's Limestone Quarry at Mandata is located on these limestone layers. The Marcellus Shale is well known today to be the target of the natural gas drilling industry. This Devonian age black shale is present within our area of study. The shale is exposed at the surface in the center of the Hoofland Valley and dips south below the Mahantongo and Hegins-Lykens Valleys. In the northern portion of the Mahantongo, it lies between 1000 and 10,000 feet below the Earth's surface. The shale is 13,000 feet below Klingerstown and 15,000 feet below the Hegins -Lykens Valley.

We will now begin our imaginary journey and leave the Hoofland Valley by walking south across Fisher's Ridge. Fisher's Ridge forms the southern boundary of the Hoofland Valley. This is the next youngest layer of sandstone and contains larger shells, and marine plant life, such as sea weed called arthrophycus.

As we continue our walk into the Mahantongo Valley we see the fields are underlain by red and yellow shale layers. As we walk south across this valley we can find shells, and crinoid marine life very easily into the center of the Valley.

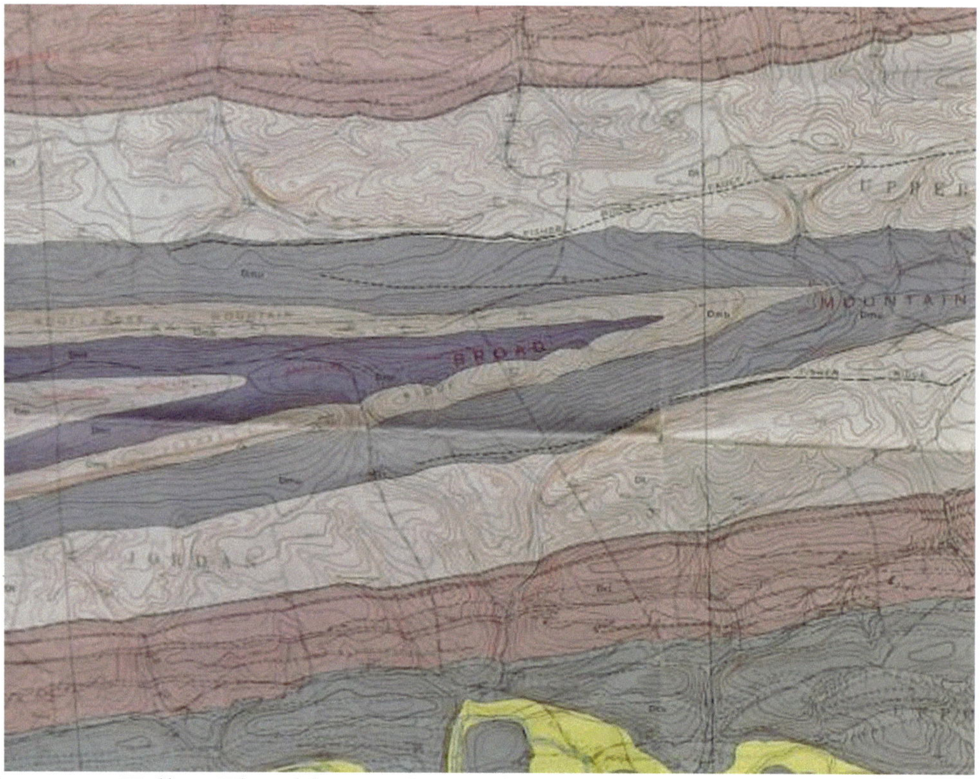

Geologic Map of the Mahantongo Valley. The oldest rocks are found in the center of the valley and the younger rocks form the Line Mountain and Mahantongo Mountain, on the top and bottom of the photo. USGS Geologic Quadrangle: 700 Trexler and Wood, G.H., 1968 see bibliography for complete source.

But as we walk further south we begin to notice less marine fossils and suddenly plant fossils are found. Small thin fossils of grasses and their rootlets which grew in the red muds can be seen. If you look you can find mud cracks, and rain drop impressions also. Geologists would term this as a terrestrial origin of these sediments. These sediments were exposed to the air sometimes. They were above the sea water periodically.

Next we come to the big Mahantongo Mountain. These sandstone and conglomerate rock layers contain a thin coal seam. This is the first coal to be formed. Plant life has now become prolific to the extent that it was compressed into a carbon layer. This layer is usually 6 inches thick or less and it is the first of many such layers to later form as plant life evolves more completely. Efforts to mine this layer of coal on Mahantongo Mountain can be seen in pits dug near the Village of Erdman in the Klingerstown Gap, as well as near Pillow on the east side

The Pillow Gap in Mahantongo Mountain is in the Pocono Formation of Mississippian Age sandstone and conglomerate. Mahantongo Formation in the foreground is of Devonian Age.

Mahantongo Mountain at Malta: The cove is created as the ridge curves to become Berry Mountain south of Millersburg.

of the Gap, near the top of the mountain. A pit was also developed on top of Mahantongo Mountain nearby the Gessner Hunting Lodge.

As we cross the Hegins- Lykens Valley we continue to have red shale. No more sea shells are found in the rock. We do find ripple marks made by shallow streams flowing across these early mud flats now turned into shale. The only fossils present are of plant life. Each step across the Hegins-Lykens Valley brings us closer to Gratz. Gratz is located on Wildcat Ridge. Wildcat Ridge is a crest of sandy shale. The presence of the sand in the shale adds resistance to the erosion of the surrounding shale layers and the sandy shale remains higher in the flat valley.

As we look further south from Wildcat Ridge we see the Coal or Thick Mountain in front of us. We all know that this mountain has coal and lots of it. Layers of coal and sandstone are on top of more layers of coal and sandstone. The environment of deposition of these sedimentary rocks has changed from deep water to shallow water and has become a swamp. Hot and humid climate conditions existed during this time to promote the growth of trees and vegetation

The Village of Gratz is located on a sandy red shale rise in the Lyken's Valley. This is the last layer of Mississippian Age red shale before the Pennsylvanian Age. The Pottsville Formation of the Pennsylvanian Age creates the Coal or Thick Mountain.

to the extent that the decomposing vegetation matter was eventually compressed by its own weight into layers of peat which would become coal. Later the pressure of the folding of these sediments changed the soft coal into our anthracite hard coal that our local miners are so familiar with.

The geologic age of bedrock also decreases in age to the north from Hoofland Valley to Mahanoy Valley. If we were to walk in a northerly direction from the center of Hoofland Valley, over Hoofland Mountain, across the Schwaben Creek Valley, over the Line Mountain, through the Mahanoy Valley to the Mahanoy Mountain, near Hunter Station, we would be crossing sedimentary layers equivalent to the shales and sandstones south of the Hoofland Valley. The Hoofland Mountain strata is the same as Fisher's Ridge. The Schwaben Creek Valley strata is the same that underlies the Mahantongo Valley. The Line Mountain rock strata is the same as Mahantongo Mountain bedrock. The Mahanoy Valley red shale is the same as the Lykens Valley red shale. The coal bearing Mahanoy Mountain has the same coal veins that are found in Coal or Thick Mountain.

Ridges and Valleys Resulting from Cycles of Mountain Building and Erosion

Now let us consider the mechanism by which these layers of sedimentary rock were deposited on top of one another in an ancient ocean basin and later compressed by plate tectonics into the folded Ridge and Valley Province we are part of today. The long ridges within our area of study trend NE-SW, with the exception of the Line Mountain which trends E-W. (See map, p. 3) These ridges are formed of sandstone and the valleys are underlain by softer shale. These layers were originally flat but were uplifted to form a mountain. This mountain was then eroded down to expose the ridges and valleys.

These ridges and valleys are formed of shale, sandstone and conglomerate layers which were formed during the Silurian, Devonian, Mississippian, and Pennsylvanian Periods which began 400 million years ago. (See Geologic Time Chart.) These sedimentary rocks were originally flat. They were formed in an ocean basin. This layering of sediment is the result of depositional cycles. As time passed, the layers were uplifted to form mountains as the Earth's crust became changed by the plate tectonics. Plate tectonics is the study of the movement of many crustal plates which together create the Earth's crust. This uplifting mountain forming event is termed the Alleghenyian Orogeny of the Permian Period. This occurred 250-290 million years ago. The Permian Period is the last of the Paleozoic Era. (See Geologic Time Scale Charts p. 9 and 11.) The geologic activity affecting PA is described for each period .

Erosion of these rocks began already at the time of beginning of the uplift. Erosion continued as the mountains rose in elevation above sea level. The soft shale layers became topographically lower than the adjacent sandstone layers. The valleys began to take their shape between the ridges.

Erosion continued at an increasing rate during the geologic periods named Triassic, Jurassic, Cretaceous, Tertiary, and Quaternary. (See Geologic Time Scale Chart, p. 11) Each of these periods describes the geologic activity affecting Pennsylvania. Erosion of the rocks of our area has been a continuing process up to the present day. The end result of this 250 million years of erosion of ancient mountains are the ridges and valleys which dominate our landscape today.

If we view the Mahanoy, Mahantongo, and Lykens Valleys, from a high flying airplane, we would see the long ridges forming our valleys and we would also see these ridges curve and close the valleys. The sandstone ridges form a zigzag pattern as they crisscross central Pennsylvania. There are six sandstone mountains within our area study but they are formed by only 3 strata of rock. (See map, page 3.)

The Devonian age Montebello sandstone of the Mahantongo Formation creates Hoofland Mountain and Fisher's Ridge. The formal name of the geologic formation

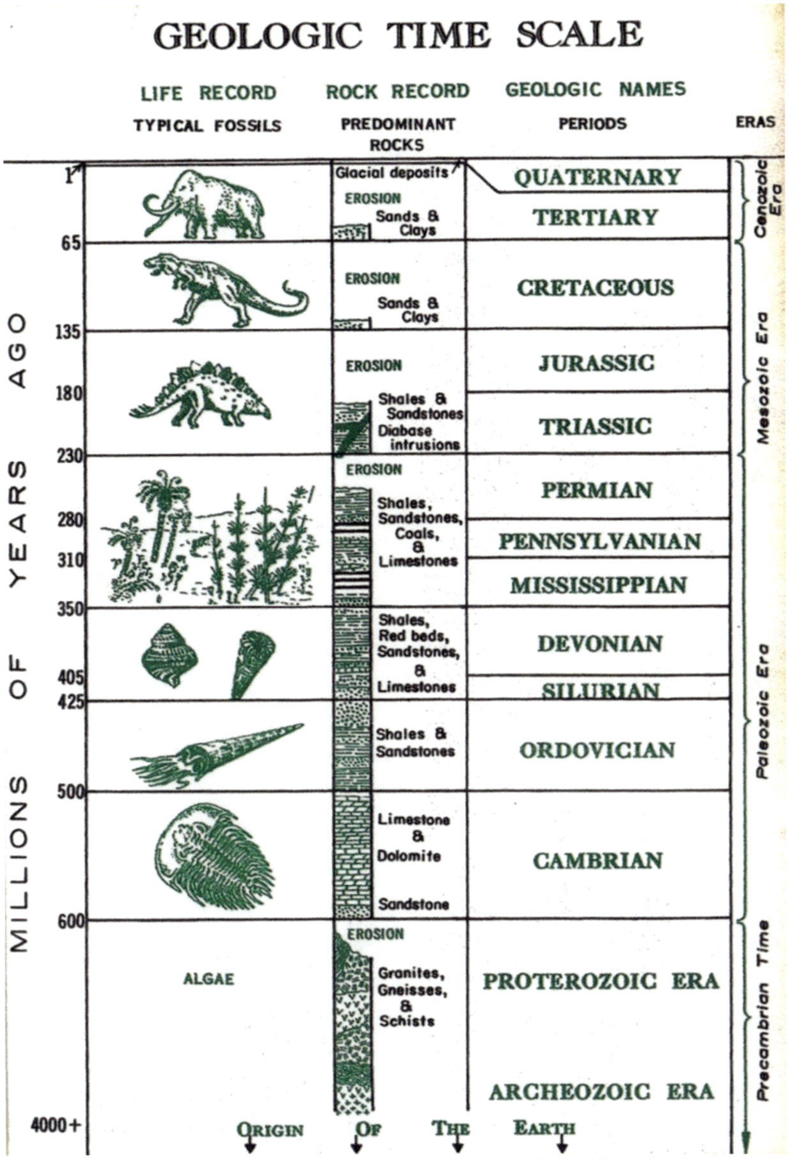

Pennsylvania Geology Summarized, back cover, Educational Series No. 4, by Bradford Willard. For additional time scale charts see the following: Pennsylvania Geological Survey; Barnes, J.H. and Sevon, W.D., 2002, The geological story of Pennsylvania (3rd ed.): Pennsylvania Geological Survey, 4th ser., Educational Series 4, 44p. Another time scale reference is USGS Fact Sheet 2010-3059. U.S. Geological Survey Geologic Names Committee, 2010, Divisions of geologic time- major chronostratigraphic and geochronologic units: U.S. Geological Survey Fact Sheet 2010-3059, 2p. http://pubs.usgs.gov/fs/2010/3059/

is Mahantango Formation. Here in the local spelling is used.

The Mississippian age Pocono sandstone–conglomerate creates Little Mountain, Line Mountain, Berry Mountain, and Mahantongo Mountain.

The Pennsylvanian age Pottsville conglomerate creates the Coal or Thick Mountain and the Big Lick Mountain. They join near Loyalton where the mountain is named Short Mountain. These ridges enclose Bear Valley. The Pottsville Conglomerate creates Big Mountain at Trevorton which joins the Mahanoy Mountain at Hunter Station.

Take note of the fact that each different period in our local geologic history is associated with a sandstone ridge forming layer. Also, each sandstone layer is separated by softer shale rock layers which underlie our valleys. There is a cycle here. Sand and mud ... sand and mud ... sand and mud. Geologists explain this cyclic layering of sedimentary rock by the simple process of varying the ocean level. We can all visualize the presence of a sandy beach at the edge of an ocean. We can all visualize the mud delta at the mouth of the Mississippi River, reaching out into the Gulf of Mexico.

An ocean level rise will place the sandy beach farther inland. The silts and mud of the deeper ocean water will migrate further inland as well, and will be deposited on top of the sand layers of the old beach.

An ocean level lowering will allow the beach sands to follow the ocean edge. The sand of the beach will cover what had been layers of silt and mud previously deposited in the deeper ocean water. Therefore the mechanism of producing alternating layers of sandstone and shale strata in an ocean basin can be attributed to fluctuating ocean levels. People who live along the coast today are all concerned about the rising ocean levels. This has happened many times before.

Visual Demonstration of Parallel Ridges

For a visual demonstration of how a flat layer of rock strata can be folded up and eroded down to form parallel ridges-fold a sheet of paper into an arch to represent a mountain. Use a scissors to cut off the top of the paper. This horizontal cut represents the erosion that has occurred over millions of years. The edges of the paper remaining represent the sandstone strata forming two parallel ridges similar to those of our area.

Visual Demonstration of Curved Ridges in Anticlines

But the ridges of our area do not remain parallel. These ridges join in coves or kettles to encircle the valleys. The following description is given to demonstrate how flat strata of sandstone can be folded and eroded lower in elevation resulting in a curving of rock ridge. For a visual demonstration of the folding of rock layers into an arch or what is called an anticline, take a piece of paper and, with a pencil, draw a line from one side of the paper to the other side. Fold the paper to form an arch with the pencil line running along the crest of the arch. The pencil line is the trace of the axis of the arch of the fold. Tilt or “plunge” the axis of this fold. Use a scissors to horizontally cut off the top of the folded paper, like the weathering which has eroded a folded mountain. The paper edge remaining represents the trace of the rock strata and will be curved just like the ridges which form the kettle of the Mahantongo Valley and Snowdale of the Hoofland Valley. The kettle is in the eastern Mahantongo Valley and Snowdale is in the eastern part of the Hoofland Valley. These ridges are formed of rock strata that is folded upward and tilted gently East and eroded so that only the roots of the strata remain.

Visual Demonstration of Curved Mountains in Synclines

The following description is given to demonstrate how coves are formed by a dip in the rock strata which is called a syncline. For a visual demonstration of the formation of a syncline –take a piece of paper and, with a pencil, draw a line from one side of the paper to the other side. Fold the paper to form a syncline with the pencil line running along the base of the down folded paper. The pencil line is the trace of the axis of the downfolded syncline. Plunge the axis of this fold. Use a scissors to horizontally cut off the top of the folded paper, like the weathering which has eroded the sandstone strata. The paper edge remaining will represent

GEOLOGIC TIME SCALE

YEARS AGO	ERA OR EON	PERIOD	ACTIVITY AFFECTING PENNSYLVANIA	MAIN ROCK TYPES OR DEPOSITS IN PENNSYLVANIA	DOMINANT LIFE FORMS IN PENNSYLVANIA
0 to 2 million	CENOZOIC ERA	QUATERNARY	Glaciation; periglacial erosion and deposition	Sand, silt, clay, gravel	Mammals, including humans
2 million to 67 million	CENOZOIC ERA	TERTIARY	Weathering and erosion; creation of present landscape	Sand, silt, gravel	Mammals, grasses
67 million to 140 million	MESOZOIC ERA	CRETACEOUS	Erosion and weathering	Clay, sand	Dinosaurs, mammals, birds
140 million to 208 million	MESOZOIC ERA	JURASSIC	Diabase intrusions; opening of Atlantic Ocean	Diabase	Dinosaurs, mammals, birds
208 million to 250 million	MESOZOIC ERA	TRIASSIC	Separation of North America from Africa; sedimentation in rift valley	Shale, sandstone, diabase	Dinosaurs, early mammals and birds
250 million to 290 million	PALEOZOIC ERA	PERMIAN	ALLEGHANIAN OROGENY: Collision of Africa and North America; mountain building, thrust faulting, and folding; much erosion	Sandstone, shale	Insects, amphibians, reptiles
290 million to 365 million	PALEOZOIC ERA	PENNSYLVANIAN AND MISSISSIPPIAN (Carboniferous)	Alluvial deposition; eastward advance of shoreline followed by development of low, flat alluvial plain	Sandstone, siltstone, shale, coal, limestone	Trees, ferns, amphibians, air-breathing molluscs, insects
365 million to 405 million	PALEOZOIC ERA	DEVONIAN	ACADIAN OROGENY: Collision of Avalonia, Europe, and North America; formation of Catskill Delta	Conglomerate, sandstone, shale	Fish, amphibians, insects, land plants
405 million to 430 million	PALEOZOIC ERA	SILURIAN	Erosion of mountains; deposition of sand and mud	Conglomerate, sandstone, limestone	Corals, fish
430 million to 500 million	PALEOZOIC ERA	ORDOVICIAN	TACONIC OROGENY: Thrusting of volcanic arc; development of Appalachian basin	Shale, limestone, dolomite	Molluscs, bryozoa, graptolites
500 million to 570 million	PALEOZOIC ERA	CAMBRIAN	Transgression of the sea; carbonate deposition	Limestone, dolomite, quartzite	Trilobites, brachiopods
570 million to 2.5 billion	PROTEROZOIC EON		Accretion of microplates to form Laurentia	Schist, slate, marble	Blue-green algae, jellyfish, worms
2.5 billion to 3.8 billion	ARCHEAN EON		Bombardment by meteorites; creation of continental crust	None identified	Bacteria
3.8 billion to 4.5 billion	PRE-ARCHEAN EON		Formation of Earth and solar system	None identified	None identified

The Geological Study of Pennsylvania, Barnes and Sevon (2002) the trace of the rock strata and will be curved just like the ridges which form Bear Valley. The cove where the Bear Valley ridges join is toward the west of the valley. This is the nose of the syncline. We refer to this nose as Short Mountain near Loyalton. Similarly, Mahanoy Mountain at Hunters Station is the nose of a syncline. Little Mountain and Line Mountain join to form the Dornsife Kettle. The Herndon Head is the nose of this syncline that creates the kettle. These ridges are formed of rock strata that is folded downward and tilted gently east.

The Geologic Time Scale

The geologic time scale describes the different Periods in the geologic history of the Earth beginning 4 billion years ago and ending in very recent time 1 million years ago. The history of our Appalachian Mountains begins only 600 million years ago. There are some older rocks nearby in the Lebanon Valley but they are not represented in our local area study. The geologic Eras of the Late Paleozoic, Mesozoic and Cenozoic will concern us. (See Geologic Time Scale Charts, page 9 and 11)

In the Paleozoic Era the Earth's surface was much different than it is today. Southeastern Pennsylvania was the border of an ancient mountain range named Appalachia. It was formed of Precambrian rocks. A large sea existed across most of Pennsylvania to where western PA is today. This inland sea extended into the interior of North America. Along the coast of this sea were sandy beaches. Limy muds washed into this sea and settled on the sea floor.

Silurian and Devonian Period

In Silurian and Devonian Period time, 400 million years ago, these sediments continued to be washed into this ancient sea. The weight of the accumulating sediments caused the sea floor to sink deeper into the Earth. Eventually the sediments became several miles thick. Most of these sandstones, shales, and occasional limestone layer formed in shallow water. This was a quiet time in geologic history. Ancient sea life became prolific as evident by all the types of marine fossils found in the Silurian and Devonian aged rocks in our area.

In the Silurian Period, limestones were deposited in the deep water of the ancient ocean basin. These limestone layers have become very important to us. The hard stone is crushed into many sizes and is used to build roads. The stone is also crushed into powder to create lime for spreading on the farmers' fields. Soil that is derived from red shale layers needs to be "sweetened" by lime. The local farmers of yesterday required lime also but modern stone crushing machinery did not exist. A different method was used to create powdered lime. The stone was burned in a lime kiln. The kiln was filled from the open top. Alternating layers of wood and stone were stacked in the kiln. The burning wood created heat necessary to break down the hard stone into lime. The powdered lime was tapped out at the bottom of the kiln.

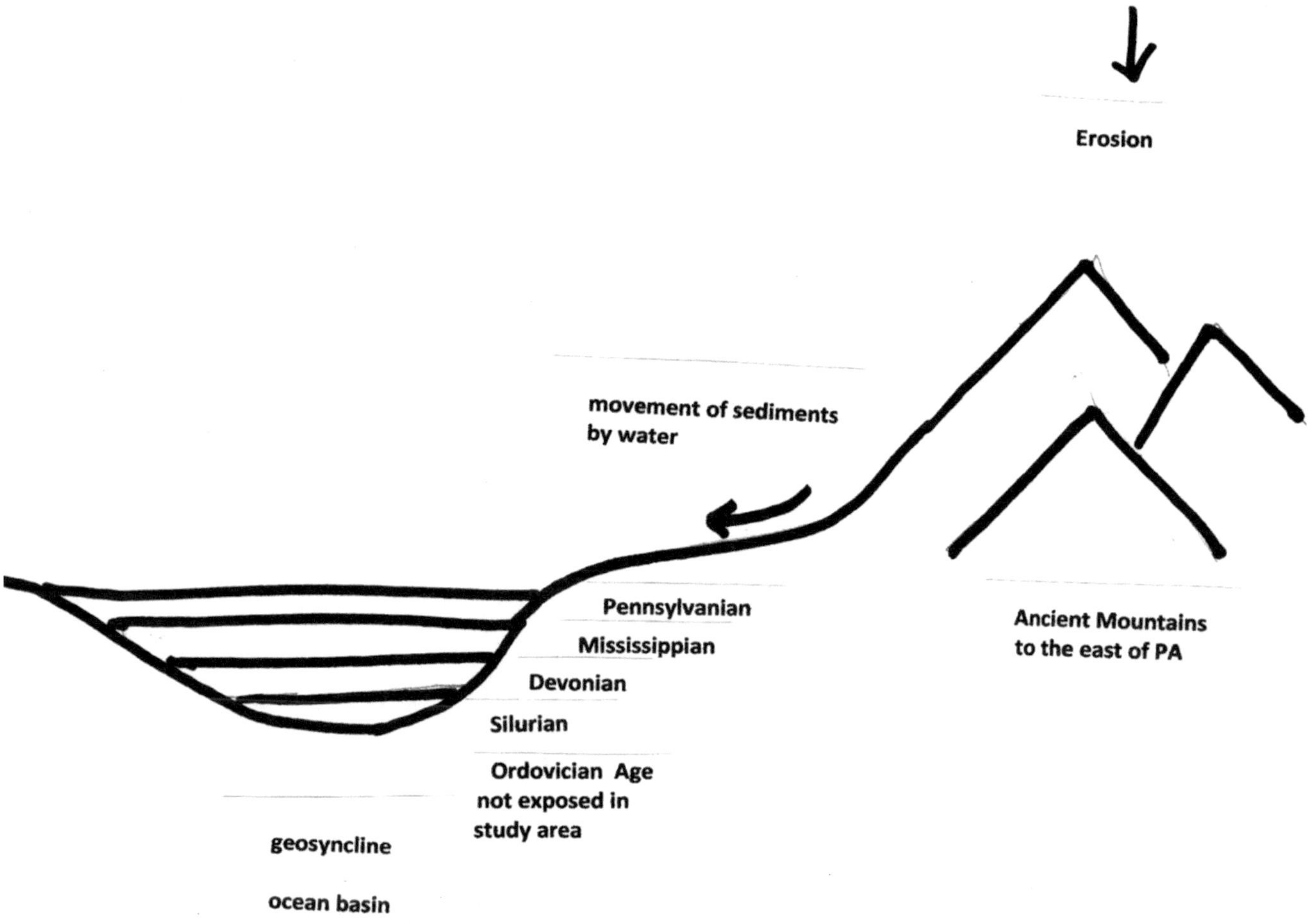

The Origin of our Local Sedimentary Rocks: The earth under our feet and the rocks we walk on in the Mahantango, Mahanoy and Lykens Valleys were formed in association with water. Here we see an ancient mountain range being eroded by natural forces including sunshine, wind and rain, freeze and thaw, into gravel, sand, and mud. Rivers and streams provided movement of these sedimentary materials into this basin. This basin is called a geosyncline, which is a depression in the Earth's crust. The oldest sediments transported into the basin are deposited on the bottom. As the geosyncline becomes filled, the oldest layers at the bottom are of limestone. Sandstone and shale, in alternating fashion, filled the basin to the top. This would have been near sea level at that time.

The Devonian Age is characterized by the Acadian Orogeny. This ancient mountain range existed to the east of Pennsylvania. (See the figure above.) These mountains were the source of the Devonian Age sediments deposited in the ocean basin that existed within our area of study. Of special interest to the author is the possible presence of volcanic ash beds within the rock layers of the Devonian Age. These ash beds have not been found locally at the present time. In USGS

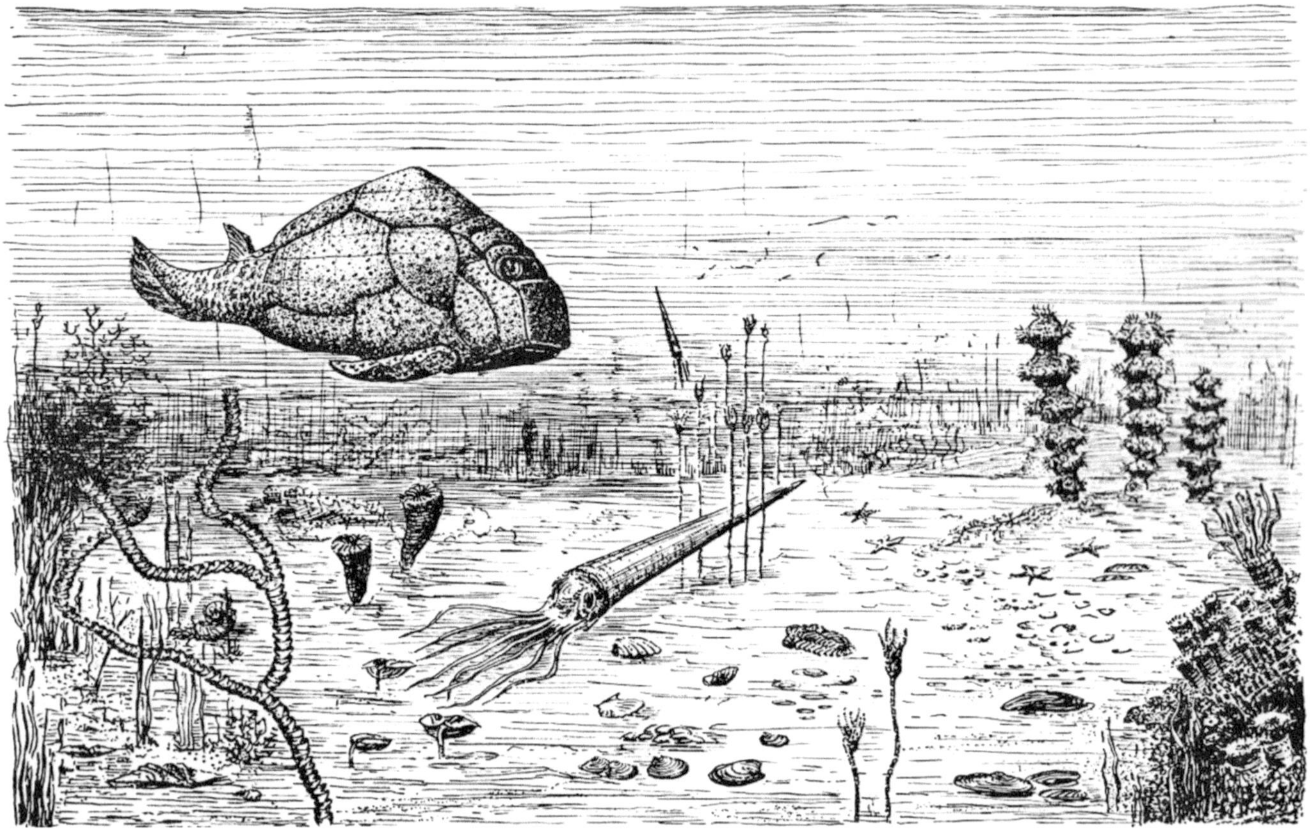

A reconstruction of a Devonian sea bottom including sponges, corals, brachiopods, pelecypods, cephaolopods, trilobite, crinoids, starfish and armored fish. Hoskins, D. M., 1969, Fossil Collecting in Pennsylvania: Pennsylvania Geological Survey, 4th Series, General Geology Report 40, NN p.4

Professional Paper 1298 (see http://pubs.er.usgs.gov/publication/pp1298), figure 4 of that report, there is mention of an ash bed at the base of the Marcellus shale. There is also mention of an ash bed above the Montebello Sandstone of the Mahantongo Formation. The Montebello Sandstone forms Fishers Ridge and Hoofland Mountain. This research courtesy of Kevin Breen.

The Catskill Formation is of Devonian Age and is exposed within the Hoofland Valley in the area known as Snowdale. In 2002, Geologist Kevin Breen, identified a mineral found during construction of a fish pond along the Snow Creek. A bulldozer uncovered shiny blue rocks that sparkled in the sunlight. This dense shiny rock was not typical of our red shale valleys. Kevin examined the rock and it was found to contain pyrolusite (MNO2). He explained “this is fairly common in the Catskill Formation and is thought to result from crystallization along fracture traces.” Fracture traces may be identified on topographic maps. They are usually linear in nature. These traces are seen on the surface of the Earth as a result of breaks in rocks due to intense folding or faulting. The headwaters of the Snow Creek flow parallel to a known fault in the bedrock of Snowdale. Movement along this fault resulted in the formation of pyrolusite.

The Mandata Gap is named for an Indian maiden who lived there. The west end of the Hoofland Valley has the oldest rocks of our area, including limestone.

An old lime stone quarry near the Village of Urban, last used by Arthur Long to demonstrate mining by man power and horse power.

Arthur Long's lime kiln between Urban and Mandata.

There are other mineralized rock layers exposed in the Mahantongo Valley from the Devonian Age. East of Snowdale, is a valley with a small stream named the Leck Bush. Heavy minerals from the Trimmer's Rock Formation were found in this stream at the base of a hill. This discovery attracted gold prospectors over a century ago. Conversation with Gene Kehler, the property owner today, revealed important information. Samuel Mattern was born in 1885 and worked in the mine. Lawrence Mattern, Samuel's grandson, now deceased, recalled high quality clay was produced. Gene Kehler's son recently examined the mine ruin and reported timbers can be seen inside the mine as well as a pool of water at the end of the tunnel. Here a slope extends downward to the east, which is now full of water. The slope remains preserved. There was very little, if any, gold found. Gene has built a new home on top of the mine hill and has a drilled water well. The water well is high in minerals including iron and sulfur. For more information on this gold mine in the Leck Bush near Rough and Ready, see the book, *There is Something about Rough and Ready.*

One of the most eccentric characters of the Mahantongo Valley was Col. George Shroeder. He was born in 1816. He laid out the town of Franklin Square. At the crossroads of this town, he dug for gold to a depth of 90 feet. The mine shaft was

PLATE II

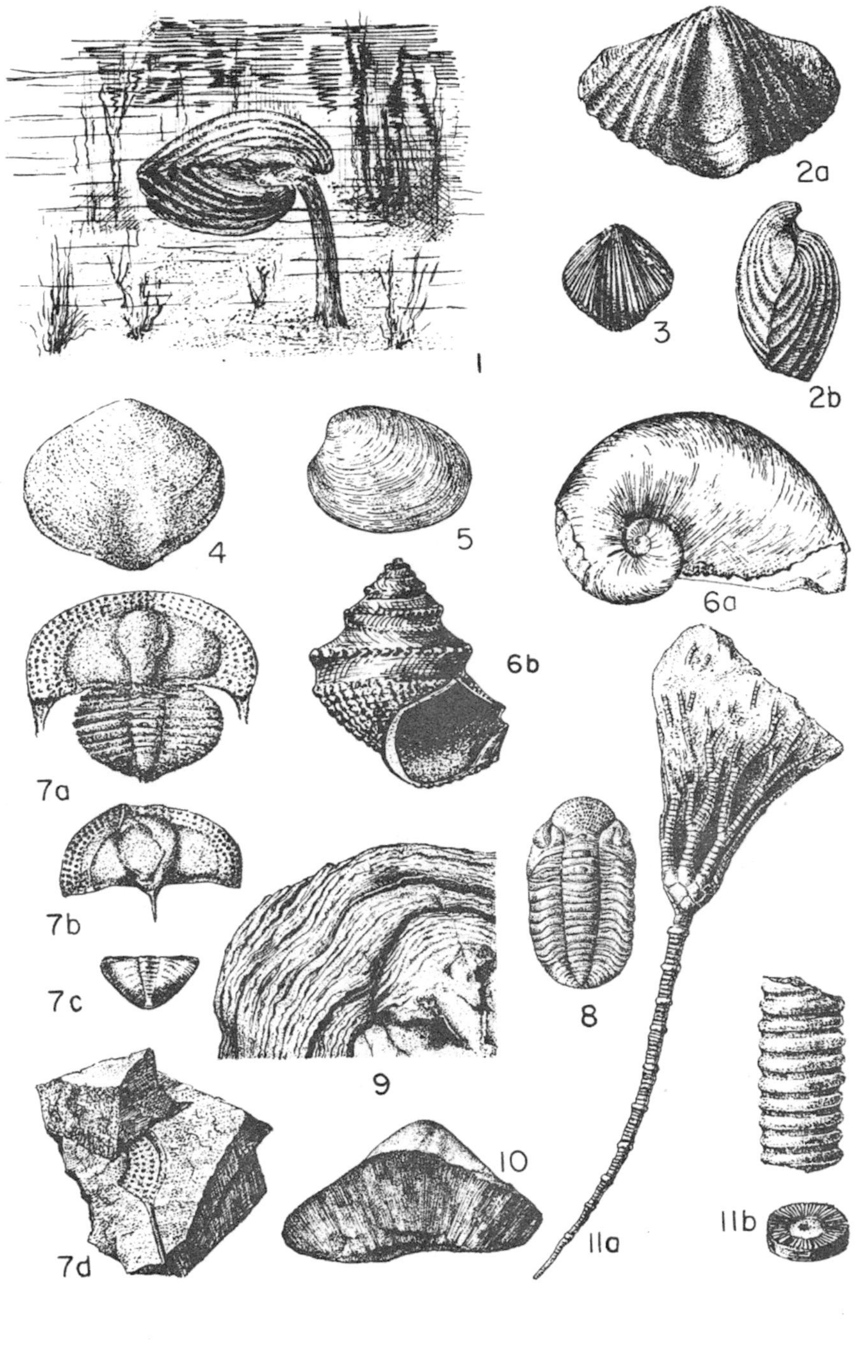

Devonian Marine Life, Common Fossils *Hoskins, D.M., 1999, Common Fossils of Pennsylvania (2nd Edition): Pennsylvania Geological Survey, 4th Series, Educational Series, 18p.*

Red Shale Hills of Rough and Ready: Mahantongo Formation transitional from marine to terrestrial origin.

8 feet in diameter. He found no gold but the hole in the earth became a well of never failing water which was used by the entire neighborhood during dry seasons. Local mining lore suggests that a tunnel was excavated at the base of the 90 foot shaft. The tunnel extended in the direction of the hotel on the opposite side of the intersection. The tunnel extended below the old hotel now known as "Jack's Spot." In the Village of Franklin Square a landmark is located. It is a purple pump. It is an old fashioned hand operated water pump on top of the gold mine. The rock excavated was from the Catskill Formation.

Mississippian and Pennsylvanian Period

In Mississippian and Pennsylvanian Period time, 350 million years ago, more sand and coarser grained sediments and large amounts of all types of sediments filled the ocean basin to its surface. The coarse grained sands of this time formed the thick sandstone layers which are the bedrock of our mountains here in Central PA. The Mississippian Age, Pocono Formation, underlies many of the

Devonian Forest Early land plants: Devonian forest plants had covered the land by the middle of the Paleozoic era, 320 million years ago. ID# CK49TCharles R. Knight paintings from the Photo Archives. All images shown in the gallery are © The Field Museum. Photograph by Ron Testa or John Weinstein.

mountains within our area of study, including Peters, Broad, Berry, Mahantongo, Line, and Little Mountains. In the Geological Survey Professional Paper 602, by Wood, Trexler, and Kehn, p. 62-65, these mountains are described as "steep, rough, and talus covered featureless slopes. Pocono rocks seem to have been accumulated principally as flood- plain deposits, to a lesser extent as swamp deposits, and perhaps locally, as lacustrine (windblown) deposits." This report describes the conglomerate rock units and the size of quarts pebbles found within these layers. This interesting study of quartz pebbles illustrates the source of the Pocono Formation as the ancient Acadian Mountains which existed in the Mississippian Age, 40 to 50 miles southeast of our area study. (See Geologic Time Scale page 11)

"In the northwestern part of the Klingerstown Quadrangle (GQ-700), the pebbles in the rocks of the Pocono reach a maximum diameter of 1 ¼ inches. Several miles to the south east, the maximum size of the pebbles increases in the Pocono rocks near Taylorsville, in the Minersville Quadrangle, (GQ-690), the pebbles are up to 2 inches in diameter." The farther south this rock unit extends, the greater the size of the pebble becomes. The Pocono Formation includes larger pebbles and cobbles ranging in size up to 6 inches in diameter, as deposited farther south nearby the source of the origin which was the Acadian Mountains.

Geological Survey Professional Paper 602, describes the mineralogy of the Pocono Formation on page 63. (See bibliography, Wood, Trexler and Kehn). The Pocono Formation contains minerals including titanium rich Ilmenite, iron rich Magnetite, and iron rich Limonite. Your author notes that iron rich nodules can be found within these sandstone conglomerate layers. They appear as purple-red spheres, usually larger than an inch in diameter, hollowed out in the center. On a recent hike to the top of Mount Mahanoy, also known as the Herndon Head, many of these nodules were found. They are also found on Mahantongo Mountain near Klingerstown.

Iron deposits are more extensive in the Pocono formation further to the south, near Loyalton, as found on Berry Mountain. These deposits were mined on the north slope of Berry Mountain as a source of iron ore for Oak Dale Forge. Local historian, Tom Buffington of Lykens, PA, has researched this early industry. The ore was removed from trenches on the surface of the mountain slope east of the Oak Dale Forge. The gray colored sand stone contains iron oxide deposits which appear as black bands within the lighter colored sandstone layers.

According to *the Iron Manufacturers Guide to the Furnaces, Forges and Rolling Mills ... by* J. Peter Leseley, American Iron and Steel Association, entry number 156, Oakdale Forge, situated on Wiconisco creek, thirty miles north of Harrisburg, owned by David K. McClure, Oak-dale P.O., Dauphin county, built in 1830, has 5 forge fires, 2 hammers driven by water and made in 1856, 237 tons of blooms and bars. Tom Buffington states, a Bloomery is a small charcoal fired, forced air chimney that doesn't actually have the Iron get to the molten stage, but rather the Iron congeals into a gob or bloom at the bottom of the chimney. That bloom is

A reconstruction of a Pennsylvanian Period coal-forming swamp. Lepidodendron tree, sigillaria tree, calamites rush, ferns and cordaites conifers. Hoskins, D.M., 1969, Fossil collecting in Pennsylvania: Pennsylvania Geological Survey, 4th ser., General Geology Report 40, p 42.

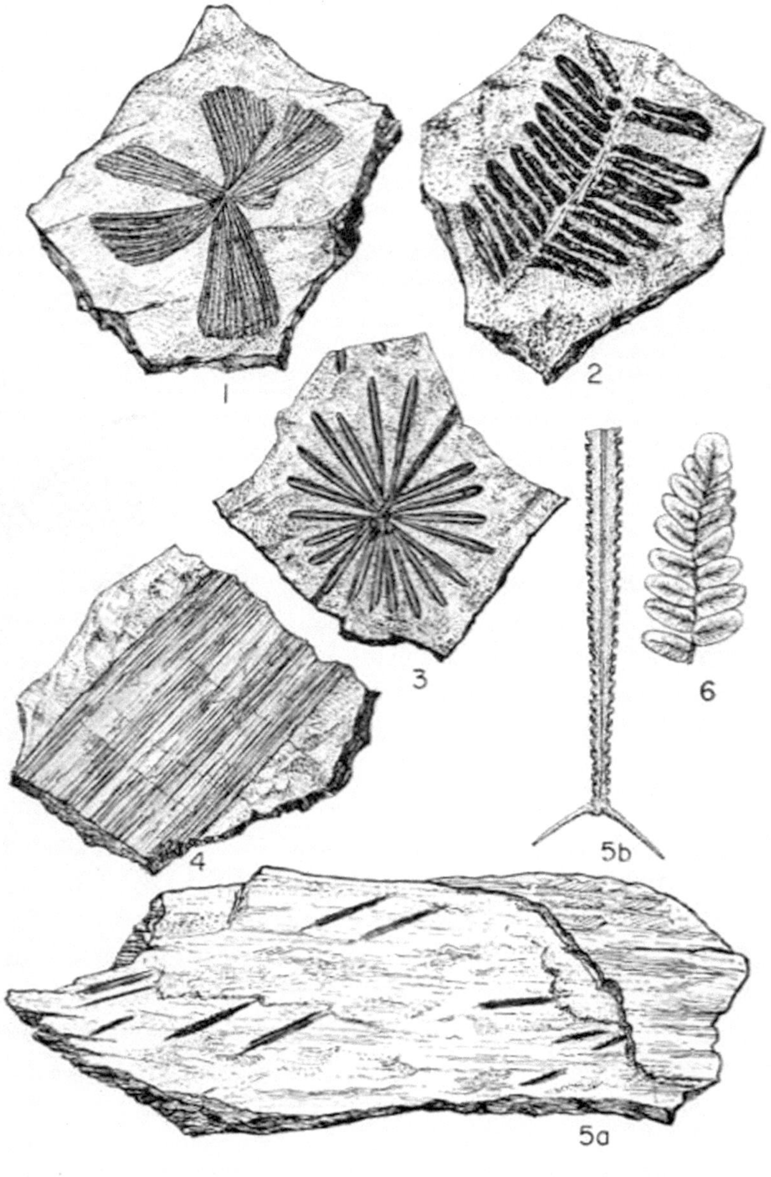

Leaf whorl, seed fern and graptolite Plate III Hoskins, D.M., 1999, Common fossils of Pennsylvania (2nd ed.): Pennsylvania Geological Survey, 4th ser., Educational Series 2, 18p.

then removed while still red hot and pounded with hammers until the end result is Wrought Iron. We also know that the Oakdale Forge was at one time owned by Dr. Charles B. McClure of Gratz, the son of David K. McClure.

By Pennsylvanian Period time the ancient sea shore was no longer across south-eastern Pennsylvania. The sea had filled up and the new sea shore was now in western Pennsylvania.

The ancient Acadian Mountains which existed to the east of Pennsylvania were more rapidly wearing down by erosion. The coal layers of the Pennsylvanian age were formed at this time on the swampy land surface as vegetation flourished. The land surface was near sea level and cycles of sedimentary layers were repeated many times. Layers of sandstones, shales, and coal were deposited over older sandstones, shales, and coal layers.

The production of a one foot thick seam of coal, requires a ten foot thick deposit of peat; so the amount of wood in the Carboniferous forest must have been immense in order to produce the hundreds of feet of coal present in Pennsylvania *(The Geological Story of Pennsylvania, p.21).*

Permian Reptiles and Amphibians: Dimetrodon, Edaphosaurus and Casea. Paleozoic era Pelycosaurs of the Permian period of Texas. ID# CK45TCharles R. Knight paintings from the Photo Archives. All images shown in the gallery are ©The Field Museum. Photograph by Ron Testa or John Weinstein.

Permian Period

In the Permian Period, 280 million years ago, big changes occurred. Continental drift collided another plate of the Earth's crust into the American land mass. It was a slow process of perhaps only inches a year or several feet per year but they collided. This mountain building event is known as the Alleghenian Orogeny. (See Geologic Time Scale Chart page 11) The sedimentary layers of the ancient sea were folded upward by this pressure and pushed into a high mountain range. These mountains covered the land of PA as we know it. The folding left some layers of rock that were raised up standing almost vertically. In our local area we often heard the coal miners talk of the steep slope of the coal veins which they followed deep down into their basin where the coal layer turned up again at a steep pitch to once more meet the Earth's surface. The folding of these layers compressed the coal into anthracite or hard coal. In western PA there was less effect from the Continental collision and the folding was much less. The layers of coal were not compressed as much and the result is a softer coal.

Plate Tectonics

The Earth's crust is formed on many crustal plates which have movement. The forces of continental drift compressed this syncline, now filled with our local sedimentary rocks. The movements of these plates is called plate tectonics.

Continental drifting compressed the flat layers of rock within the geosyncline. The oldest sedimentary layers were the first to be deposited in the deepest part of the water. These are the Silurian Age limestones.

Devonian Age shales covered the limestone. These strata are filled with many sea shells and other marine life fossils.Mississippian Age siltstones and shale covered the Devonian shales.

Mississippian Age rocks contain the earliest plant fossils found in our area. The presence of ripple marks in these strata denotes a shallow water environment. The youngest rocks on top are from the Pennsylvanian Age. Now the sandstone and coal layers alternate in a cycle deposition of sandstone and coal, sandstone and coal.

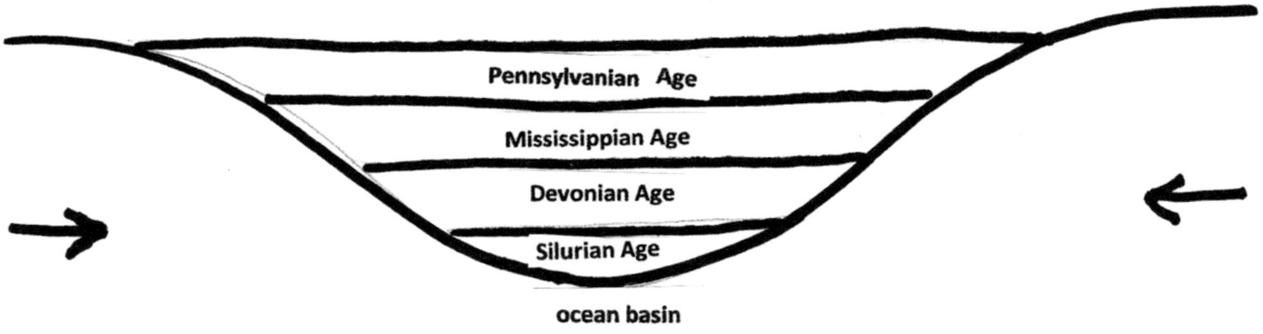

The formation of many separate coal veins was caused by the fluctuation of the ocean level. The geosyncline, now filled to near sea-level, became a shallow swamp of ancient vegetation. A tropical forest of plants and trees degraded into layers of peat that became covered by other sediments when the ocean level rose. This vegetation debris was transformed into the famous coal measures of Anthracite coal found here in Pennsylvania. When the ocean level receded a bit, vegetation again grew.

Folding of Flat Sedimentary Layers

The sedimentary rock layers of the ancient geosyncline were compressed horizontally by the forces of plate tectonics, and shortened in a horizontal direction. This forces the flat layers of sediments to be uplifted in arches called anticlines and depressed to form valleys called synclines. To visualize this, movement of the Earth's crust, picture a table cloth on a flat table. If you push on the table cloth, it will wrinkle and form a series of up and down folds. This is the

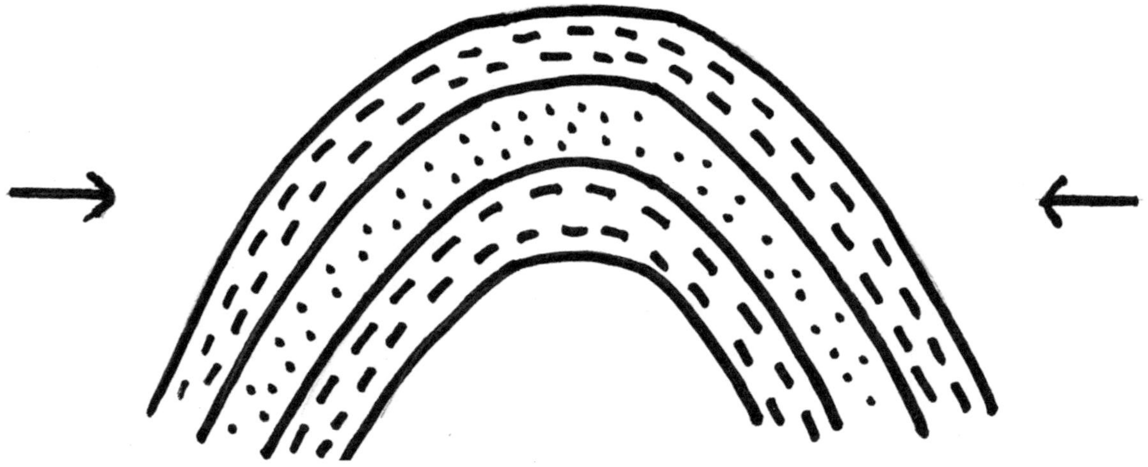

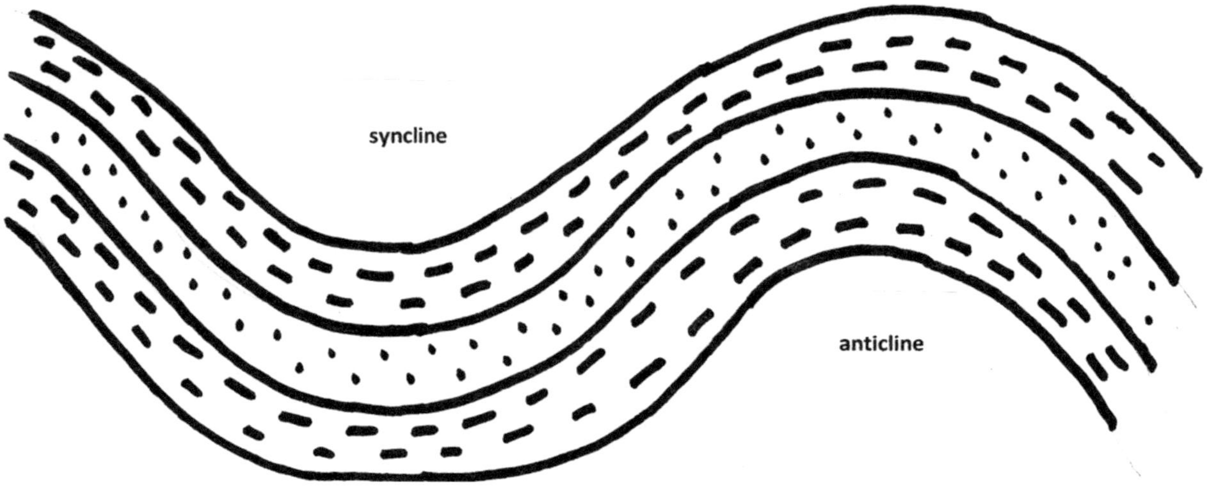

same concept of the forces which folded the geosyncline except it contained many more layers.

The Mahantongo Anticline

This anticline uplifted by plate tectonics was exposed to erosion many, many millions of years. The highest part of the folded rocks lowered by the effects of natures' elements and the mountains were eroded down to a lower level. The sandstone layers, being the more resistant to erosion, remained higher than the softer shale layers which eroded more rapidly.

As an example, here we see the Mahantongo Vallery geology. The Mahantongo Mountain and the Line Mountain stand high as sandstone ridges, and the lower valleys, underlain by shale, flank the mountains on each side. Every time it rains in our valley, the creeks further erode the underlying shale layers as evidenced by the muddy waters we see flowing in the creeks.

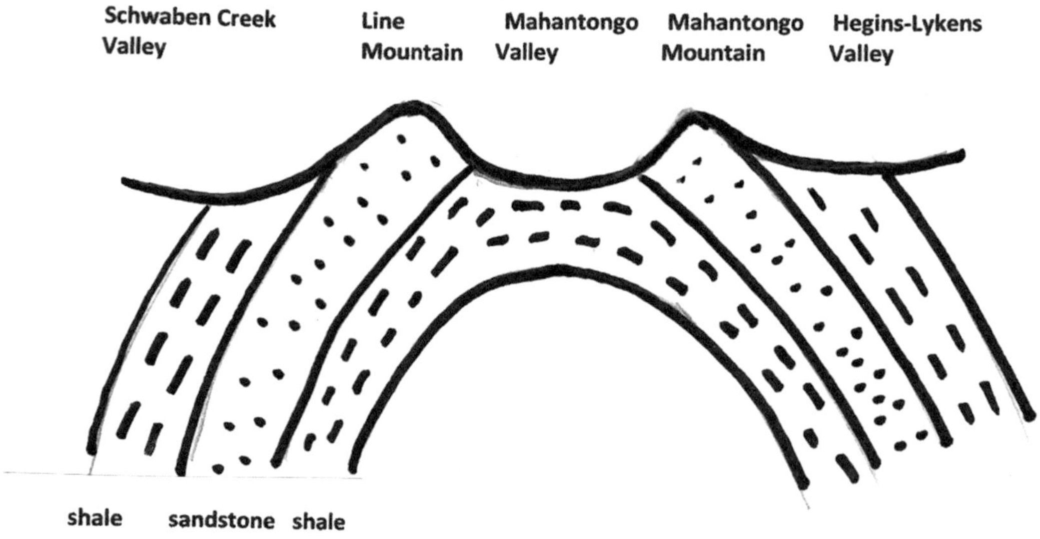

The rock strata within the Mahantongo Mountain dips to the south. The rock strata within the Line Mountain dips to the north. The axis of the folded strata trends east-west. A geology term describing the direction of the axis of the folded strata is called a strike. The strike of the rock in the Mahantongo Valley is generally east-west.

Mahantongo Mountain and Line Mountain join in "The Kettle". Note the uniform ridge top elevation marks the level of the peneplain in the age of the dinosaurs . The Pocono Formation of Mississippian Age formed into an anticline.

Klingerstown Gap - A view from The Tulpehocken Trail north of Klingerstown.

The Mahantango Anticlinerium

This illustration of the series of sedimentary rocks of our area is more detailed than the previously illustrated Mahantongo anticline. The oldest Silurian Age limestone is exposed at a quarry near the village of Mandata. Above the limestone is the Marcellus Shale, a natural gas reservoir, which is a layer within Devonian Age shales. These Devonian shales are exposed in the Hoofland Valley. A Devonian Age Montebello Sandstone layer forms Fisher's Ridge and Hoofland Mountain. Fisher's Ridge and Hoofland Mountain stand topographically higher than the valley.

Younger shale layers of Devonian and Mississippian Age, named the Catskill formation, form the Schwaben Creek Valley and the south portion of the Mahantango Valley which is drained by the Mahantango Creek.

The Pocono Formation of Mississippian Age forms the Line Mountain and the Mahantango Mountain. The Pocono Formation is sandstone and quartzite which is very resistant to erosion and remains topographically high in elevation.

Please note the this series of diagrams illustrating rock structure is simplified for visual concept. Faulting within these rock layers has contributed to geologic structure as well. Some of these faults are subterranean and not visible at the surface. The traces of some faults can be seen on the surface. The Fisher Ridge fault can be mapped on the Earth's surface in the area north of Klingerstown. The precipitous cliff eroded by the Snow Creek in the Snow Creek Valley had its origin with this break in the rock layers. The south side of the fault uplifted. The Snow Creek follows the east-west trending fault in the Snowdale Valley between the uplifted land and the Fisher's Ridge on the north.

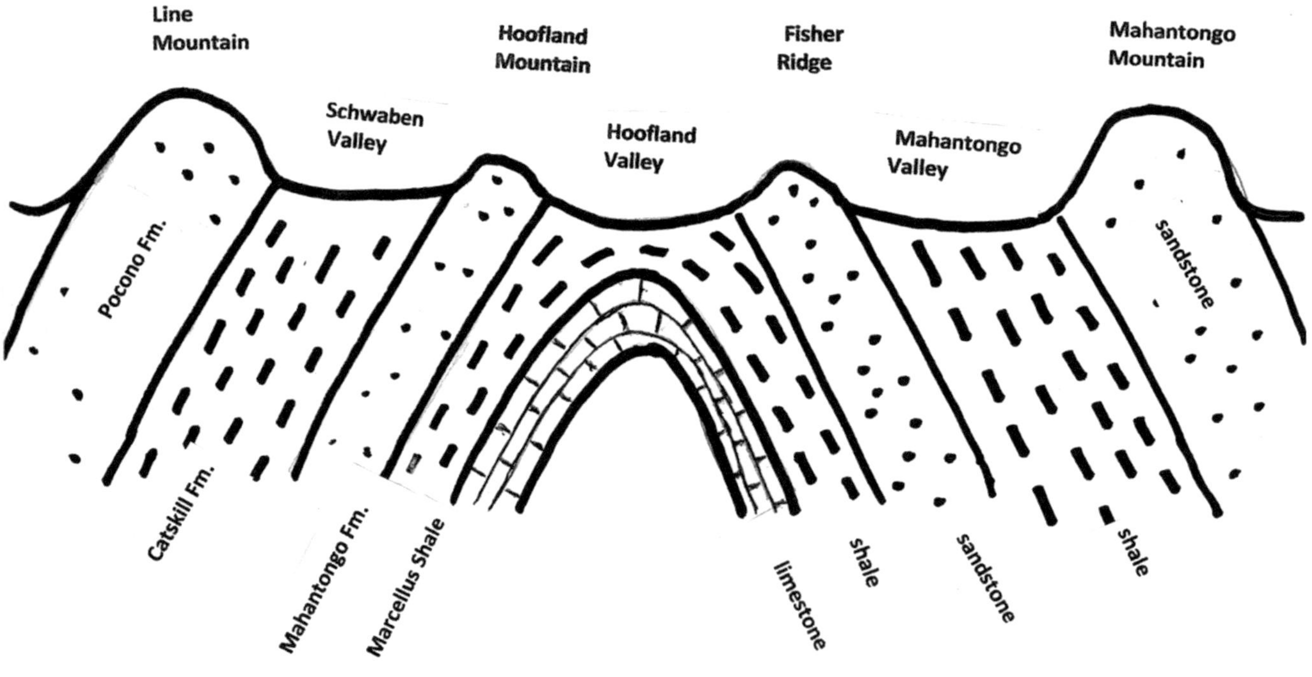

Hoofland Valley is encircled by the anticline formed by Fisher's Ridge and The Hoofland Mountain Montebello Sandstone.

The Village of Urban: Ludwig Maus was the pioneer who lived here. He named his land "The Mouse Hole."

The Mouse Creek Water Gap in the Hoofland Mountain at Urban. The mountain rock is Montebello Sandstone of Devonian Age.

Hoofland Valley- Devonian Age: Shriver Chert was mined here by Indians for stone tools.

Bear Valley Syncline

Note that this rock structure is identical to that of the Mahanoy syncline. The rock strata are the same. The down folded synclines of Bear Valley and Mahanoy Valley are twins. The mountains and valleys have different names.

Here we see Mahantongo Mountain and Berry Mountain of Mississippian Age Pocono sandstone formation. The Hegins Valley and the Lykens Valleys are underlain by Mississippian Age Mauch Chunk red shale formation.

The Bear Valley syncline is composed of younger rocks and rests on top of the Mauch Chunk and Pocono formations. Coal or Thick Mountain which is south of Gratz is formed of Pennsylvania Age Pottsville conglomerate. Big Lick Mountain, north of Williamstown, is formed of the same Pottsville conglomerate. These mountains contain anthracite coal measures. Bear Valley, famous for its coal mines, is between Coal or Thick, and Big Lick Mountains. Lykens and Williamstown are communities with a long heritage of anthracite coal mining and processing. They are located south of Bear Valley.

If a visual comparison were made between the nose of the syncline at Loyalton called Short Mountain, and the nose of the Mahanoy syncline at Hunter Station, the viewer would see the same rock structure at both locations. They look the same.

Many old coal miners state the fact that the coal bearing mountains do not reach the Susquehanna River, as many of the other ridges do. This is true. The

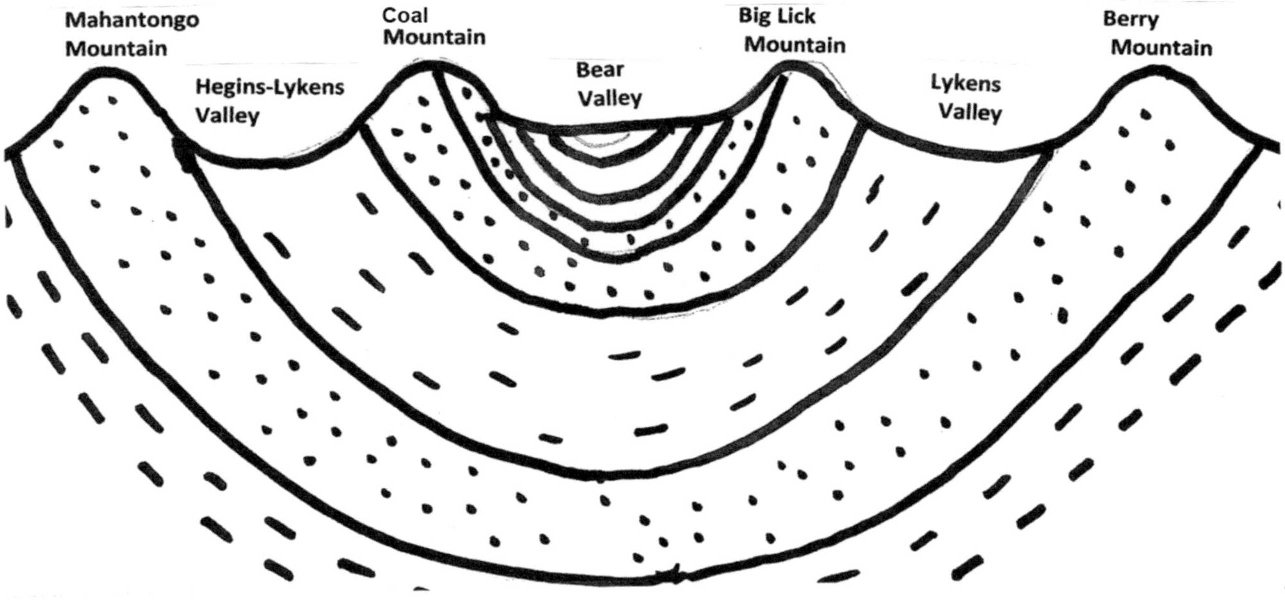

The "nose" is a geologic structure created by the plunging of the axis of the syncline. In both synclines the axis of the fold plunges gently east. Two coal bearing mountains join together to form the nose of the syncline. More about the formation of coves, kettles, and noses will be discussed in the following pages.

coal veins in the Bear Valley syncline and the Mahanoy syncline begin at a point called the "nose" of the syncline. It is a coincidence of location that in both cases, the "noses" are located east of the Susquehanna River.

Short Mountain at Loyalton is the nose of a syncline formed of Pennsylvanian age sandstone, conglomerate and coal.

A view of the cove formed by Mahantongo and Berry Mountains. The cove is across the Susquehanna River. The Lykens Valley is within the syncline of Pocono Sandstone ridges. The uniform ridge top marks the peneplain elevation during the age of the dinosaurs.

Mahanoy Syncline

This syncline formed by plate tectonics was exposed to erosion for millions of years of geologic history. The surface of the Earth was eroded down. The top of the mountain was removed by erosion. The older sandstone layers of the Line Mountain and little Mountain (Dornsife Mountain) and the younger sandstone layers of Big Mountain and Mahanoy Mountain remained at a higher elevation as ridges. The Big Mountain and Mahanoy Mountain contain coal measures.

In contrast is the Trevorton Valley formed by Zerby Run and the Cameron Valley eroded by the Mahanoy Creek. These valleys are underlain by softer red shale; more easier eroded to a much lower elevation. These red siltstone and red shale layers can be viewed from lower Road in the Cameron Valley between Gowen City and Hunter Station. The red rock visible in the road cuts is remarkable.

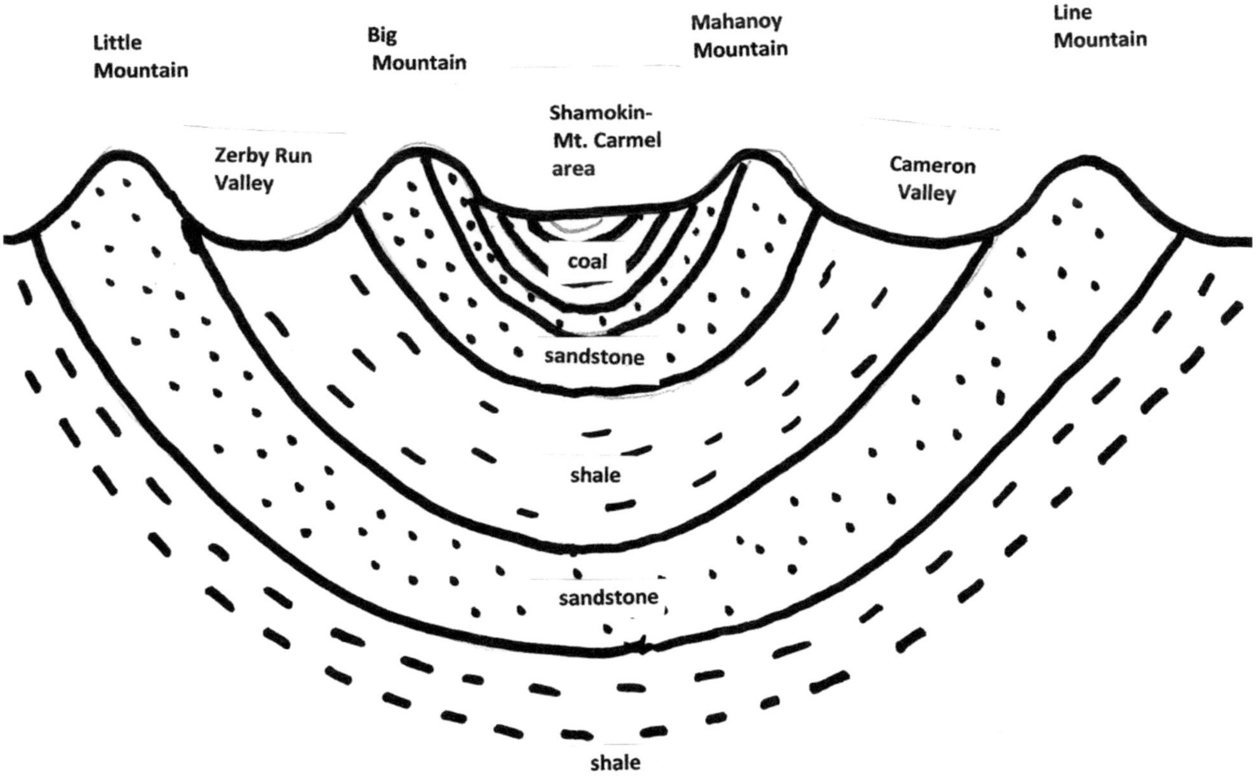

The Village of Red Cross was earlier named Mahanoy: Dornsife Gap and Herndon Head as seen from Red Cross.

The Herndon Head is the nose of a syncline composed of Mississippian Age Pocono Formation of sandstone and conglomerate.

Mahanoy Mountain-Nose of a syncline. Big Mountain and Mahanoy Mountain join here at Hunter Station.Line Mountain and Little Mountain, seen on photo edges, curve to the west to form the Kettle at Dornsife.

Mahanoy Coal Mountain at Hunter Station: This is the nose of a syncline created by PA age coal bearing rock strata.

Broad Mountain Anticlinorium

A study of the geology of our area must include a cross section of the huge anticline containing the Armstrong and Powell's Valleys. Please note that this geology is identical to that of the Mahantongo anticlinorium. The rock forming these mountains is *the* same. The mountains *have different names.*

Berry Mountain, south of Loyalton and Millersburg, and the Peter's Mountain are composed of Mississippian Age Pocono Formation sandstone. The Armstrong Valley and the Powell's Valley form the older central portion of the anticline. Devonian Age Montebello sandstone rises in the center to create a drainage divide separating the Armstrong Creek from the Powell's Valley creeks which have eroded the Catskill formation of red shale.

The valley north of Berry Mountain is the Lykens Valley containing the Village of Berrysburg and surrounding communities, underlain by younger rock strata which is the Mauch Chunk red shale formation. St. Johns Lutheran (Hill) Church is situated on a rise within these shale layers.

Big Lick Mountain is the youngest rock strata on this diagram, formed of Pennsylvanian Age Pottsville conglomerate.

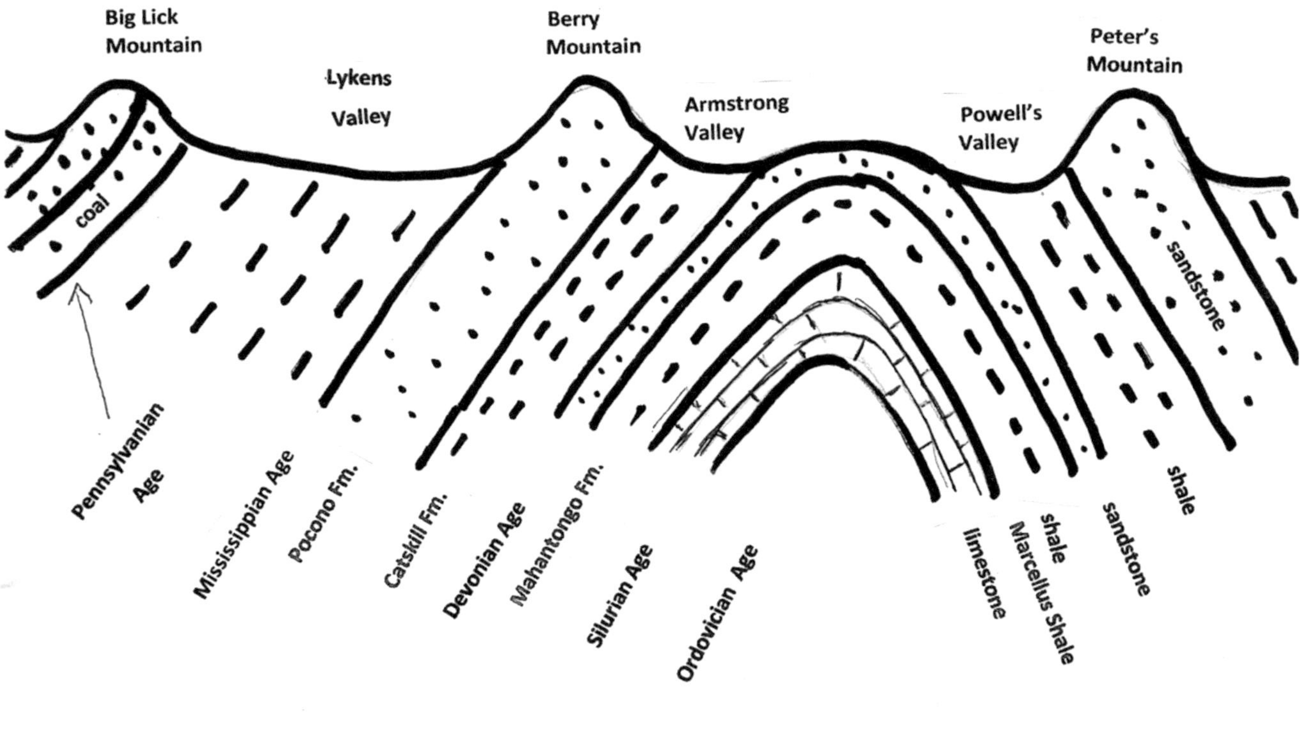

Triassic and Jurassic Period

Next was the Period in geologic time known as the Triassic and the Jurassic. These periods began 230 million years ago. Today we refer to this time as the age of the dinosaur.

The mountain building Continental drifting plates stopped moving. The new Appalachian Mountains they had pushed up began to feel the effects of earthly weathering. Water, wind, ice, heat, and cold are forces that wore down the mountains. For 100 million years this erosion continued until the land forming PA was completely eroded to become flat land called a Peneplain. All of central PA was a vast low plane of little topography. The elevation of this Peneplain is marked by the ridge tops in our area today. The Peneplain surface was above the present land surface. Ancient rivers and streams meandered slowly across the plane. This was at an elevation high above our valleys. During this Triassic Period, the Earth's crust developed cracks or faults in the bedrock. Magma from deep within the Earth's mantle pushed upward. The hot liquid was forced by pressure into the breaks in the bedrock. This hot liquid magma cooled into a rock named diabase and its shape is called a dike. We have a diabase dike in our area between Pillow and Berrysburg. Diabase iron rich rock, rusty orange in color, can be found in farm fields as what geologists call float. The diabase erodes into round cobbles. Some are as big as grapefruits and some as little as an egg. The red, yellow round

St. John's Lutheran Church, called the Hill Church, sits on a rise in the Lykens Valley formed by the Mississippian Age Mauch Chunk Red Shale. The hot diabase dike pushed upward into the shale.

View from the Hill Church: Berrysburg has a scenic horizon of the Pillow Gap. The trace of the diabase dike can be found on the surface in the direction of the gap. Diabase is igneous rock.

Anchisaurus, a Triassic dinosaur, was probably 6-7 feet in height. Hoskins, D.M., 1969, Fossil collecting in Pennsylvania: Pennsylvania Geological Survey, 4th ser., General Geology Report 40, p 40.

cobbles are easily seen when the fields are tilled and can be mapped to mark the dike's location. This is the only igneous rock type in our area as far as I know.

During the Jurassic Period, 180 million years ago, Continental drift began to be active again. But instead of crushing one continent into another, the opposite took place. The two continents previously joined together, began to separate. Portions of what are now North and South America began to separate from lands which eventually became Europe and Africa. The Atlantic Ocean began opening and this drifting apart continues yet today at the rate of 1 ½ inches per year.

The Pennsylvanian age coal fields were divided as the newly created Atlantic Ocean began to open. The Welsh people of Great Britain today mine the same coal the Pennsylvania miners do.

Archaeopteryx, Compsognathus and Rhamphorhynchus. Germany, Mesozoic, Lower Jurassic era. ID# CK39TCharles R. Knight paintings from the Photo Archives. All images shown in the gallery are © The Field Museum. Photograph by Ron Testa or John Weinstein.

Triceratops and T. rex Tyrannosaurus rex Mesozoic. ID# CK9T Charles R. Knight paintings from the Photo Archives. All images shown in the gallery are © The Field Museum. Photograph by Ron Testa or John Weinstein.

Dinornis, extinct bird. Giant Moas. Cenozoic Era. ID# CK32T Charles R. Knight paintings from the Photo Archives. All images shown in the gallery are © The Field Museum. Photograph by Ron Testa or John Weinstein.

Apatosaurus (Brontosaurus). Late Jurassic period, Mesozoic era. ID# CK5TCharles R. Knight paintings from the Photo Archives. All images shown in the gallery are © The Field Museum. Photograph by Ron Testa or John Weinstein.

Jurassic dinosaur Stegosaurus. ID# CK31TCharles R. Knight paintings from the Photo Archives. All images shown in the gallery are © The Field Museum. Photograph by Ron Testa or John Weinstein.

Cretaceous Period

In Cretaceous time, 135 million years ago, the age of the dinosaur was ending. As the movement of the Earth's plates began to open to form the Atlantic Ocean, this caused rapid erosion. The Peneplain covering Pennsylvania was affected. Water began flowing at a more rapid rate. The ancient rock layers covered by the mantle of soil of the Peneplain began to be exposed. The soil washed away. The subsoil and the loose rock covering the bedrock washed away also.

Late Cretaceous Dinosaurs, Mesozoic era. Helmet-crested Corythosaurus, a herd of long-crested Parasaurolophus, and heavily armored Palaeoscincus. ID# CK13AT. Charles R. Knight paintings from the Photo Archives. All images shown in the gallery are © The Field Museum. Photograph by Ron Testa or John Weinstein.

Great Ground Sloth, Megatherium. Nearly elephant size, and two types of armadillo like glyptodons, spiked tailed Doedicurus and Glyptodon. Charles R. Knight painting. ID# CK21T Charles R. Knight paintings from the Photo Archives. All images shown in the gallery are © The Field Museum. Photograph by Ron Testa or John Weinstein.

Cave Bear. Cenozoic era. ID# CK28T CK21T Charles R. Knight paintings from the Photo Archives. All images shown in the gallery are © The Field Museum. Photograph by Ron Testa or John Weinstein.

Tertiary Period

In the geologic period of the Tertiary, 65 million years ago, the erosion continued. A gentle uplift of the region, including PA, sped up the erosion by weather and stream waters. As the softer shales and limestone layers became lower, the sandstone and conglomerate bedrock layers remained as high lands slowly showing their well known ridge shapes. Eventually the harder rock layers remained standing high as ridges and the softer rocks were worn down into lowlands and valleys. The ancient Appalachian Mountains had been worn down to a Peneplain and now that Peneplain had been eroded to expose the roots of the same Appalachian Mountains. We have a new ridge and valley mountain range which was formed from the folded basement rock of the older mountains. The old streams and rivers which had flowed across the Peneplain slowly deepened their channels. They cut across the bedrock strata as the old drainage pattern was lowered upon the present ridge and valley topography. It was during this period that the last remaining evidence of the dinosaurs was removed by erosion. The peneplain land surface in Central Pennsylvania that the dinosaurs walked on was washed away, in addition to all the evidence of the giant reptiles such as fossil bones and the tracks of their footprints.

As the valleys became eroded lower, the meandering streams of the old Peneplain sometimes abandoned their channels or were captured by other streams. This results in the formation of a wind gap. Wind gaps like Deibler's Gap and the gap in Fisher's Ridge in Hoofland are examples of old abandoned water courses. The streams which formed these wind gaps moved to other locations as further valley erosion changed the drainage pattern.

The main streams and rivers kept their channels. They were not easily diverted. Their flowing water slowly cut across the harder rock layers and deepened the valleys. The Susquehanna River of the old Peneplain slowly cut a deeper channel as the river eroded through the Pocono Sandstone to create scenic water gaps in Mahantongo Mountain and also Berry Mountain near Millersburg. At Dalmatia the Susquehanna River cut down through the Hoofland Mountain sandstone to create McKees Half Falls. The falls are created by the bedrock Devonian strata called Montebello Sandstone. Smaller streams like the Pine Creek cut down through Mahantongo Mountain creating a water gap between the Village of Erdman and Klingerstown. Another stream today called the Deep Creek cut down through Mahantongo Mountain near Pillow creating the Pillow Gap. In Colonial times this was known as Haines Gap.

There is a gap in Fisher's Ridge which opens up the Hoofland Valley north of Klingerstown. This gap was created by a much larger current of water than the small stream located in this gap today. There is a drainage divide in the Hoofland Valley a bit west of the wind gap. This divide now controls the Hoofland Valley drainage. This drainage divide did always exist. In prehistoric time a powerful eastward flowing channel leading water from the Hoofland Valley through this gap was captured by a westward flowing stream which now drains most of the Hoofland Valley. This stream is named Mouse Creek. It passes through a water gap near the Village of Urban which is in the Hoofland Mountain. Hoofland Mountain is a high ridge of Montebello sandstone.

The end result of this erosion throughout geologic history is the land we love today. The water and wind gaps are beautiful. The flat top mountains separating our valleys create unique areas of geography. We identify with these valleys as they are the special locations of our homes. I often say that I am from the Mahantongo Valley. Other valleys are named Hoofland, Slutter, Schneedahl, Hegins, Bear Valley, Lykens, Luxemburg, Mush Valley, Raupstall, and Toad Valley. Can you name any more?

Wind Gap in Fisher's Ridge near Klingerstown: This stream was captured by the Mouse Creek during the Tertiary Period.

Pine Creek Water Gap at Klingerstown seen from Benigna's Winery. Benigna was the daughter of Count Ludwig von Zinzendorf, a Moravian Missionary. Pine Creek was once known as Benigna's Creek, or Kind Creek.

"Deibler's WInd Gap" in Mahantongo Mountain as seen from the County Line Church. During the Tertiary Period a stream abandoned its course and the formation of a complete water gap never occurred.

Deibler's Wind Gap viewed from the Hill Church. A stream abandoned its course, and what could have been a water gap was not completed.

Orohippus and Uintatherium Small four-toed horses of the genus Orohippus and the much larger, but harmless Uintatherium of the Eocene. ID # CK46TCharles R. Knight paintings from the Photo Archives. All images shown in the gallery are © The Field Museum. Photograph by Ron Testa or John Weinstein.

Quaternary Period - The Ice Age

The Quaternary Period marks the last 1 million years of Earth's history. During this period, Pennsylvania was affected by the movement of large sheets of ice and glaciers. Glaciers of ice pushed south over the land surface in advance of the ice sheet close behind. On at least 4 occasions thick ice sheets pushed down from northern Canada and Labrador. These ice sheets covered the mountains. As the ice moved across the bedrock, the heavy ice had extreme erosive power and ground down the rock. Water melting from the ice sheet moved the ground-up rock into elongate piles of gravel debris called moraines. One type of moraine which marks the stopping place of a glacier's movement is called a terminal moraine. These moraines remain as evidence of ice and water movement of gravel.

The last ice age ended 10,000 years ago. It is called the Wisconsin Ice Age. The Wisconsin Continental Ice Sheet did not reach our area. The Wisconsin terminal moraine marking the southernmost advance of ice is located near Berwick in the Susquehanna River Valley. The Illinoian Ice Age glaciers were older and larger. The Illionian ice advanced south across PA covering the land with ice several thousand feet thick. This ice advanced south, up against Little Mountain at Dornsife. This mountain is locally called the Dornsife Mountain. Although this ridge stopped the southward movement of ice, the glacier flowed around the Herndon head and pushed down the Susquehanna River Valley as a tongue of ice to the area of Paxton where it touched Mahantongo Mountain.

The ice advanced down the valley as a tongue several miles long. This tongue of ice acted as a dam to the waters of the tributary streams like the Mahanoy and Mahantongo Creeks, forcing the creeks to flow between the ice tongue and the valley wall. Large amounts of river gravel were washed away by the melting ice water as the rivers carried the glacial outwash downstream. The streams were eventually choked with these gravels and created terraces which remain high above the present water course. Louis Peltier of the PA Geological Survey was one of the first to map the terminal moraines of the Illinoian and Wisconsin glaciers in the Susquehanna Valley. In 1949, his study was published as identified in the bibliography of this text.

During the Illinoian Ice Age the glacial tongue flowed down the Susquehanna River Valley to touch the Manantongo Mountain seen here below Dalmatia.

The woolly mammoth -- a Pennsylvania resident during the Ice Age: This scene could represent the Illinoian ice advance toward Little Mountain, also known as the Dornsife Mountain. The wooly mammoths are headed south, crossing the mountain into the Mahanoy Valley of Northumberland County, PA, 75,000 years ago. This pen and ink sketch by Albert Van Olden, Pennsylvania and the Ice Age, by Vincent C. Shepps .

Glacial Topography of Rivers and Streams

Frank Leverett studied glacial deposits in PA in 1926. In Geological Survey Bulletin, G7, he states "the Susquehanna River Valley between Sunbury and Harrisburg has terraces in the lowlands between the ridges. The best defined terrace is about 40 feet above the river with a few places about 60 feet. Theses terraces belong to the Wisconsin glacial stage. The terrace at the mouth of the Mahanoy Creek above Herndon has a broad development and is 30 feet above the river. At the Mahantongo Creek the terrace is 50 feet above the river. (Leverett, p. 21) At numerous points along the Susquehanna River Valley there are rock shelves carrying deposits of gravel and sand in which granite and other erratic pebbles of glacial derivation are present. These deposits appear to be a valley train connected to Illinoian drift. They occur up to about 150 feet above the river at Fisher's Ferry. (p. 37) It is probable that the filling of the river channel with glacial gravel was up to 100 feet above the river but it is now preserved at its full height at only a few places." Mr. Leverett also suggests the movement of the Illinoian glacial tongue of ice eroded the precipitous bluff opposite the city of Northumberland which stands at the junction of the west and north branch Susquehanna River Valleys. This bluff was over steepened by the movement of the ice. Today many people view the Indian Shikellamy's Profile here. The cliff rises almost perpendicularly to a height of 400 feet above the river. (Leverett, Bulletin, G7, p. 26)

Another example of an Ice Age outwash terrace can be found near the Village of Hunter Station. East of the Emmanuel Lutheran and Reformed Church on top of the hill, gravel and round cobble stones can be found in a road cut and in nearby fields north of Route 225.

This mixture of outwash gravel (below) is high above the Mahanoy Creek's present elevation. During the ice age when the Susquehanna River was filled with outwash gravel, the river flowed at a higher elevation, therefore the tributary creeks also flowed at a higher elevation as determined by the river's elevation. The Mahanoy Creek deposited a gravel bar on the hilltop in Hunter Station when the Mahanoy Valley was partially filled with sediment and gravel up to the hilltop elevation where it was deposited by the creek. When the ice age conditions ended most of these gravels, but not all, were washed downstream as the Susquehanna River Valley eroded lower to its present elevation. Some gravel bars and terraces were left behind perched as remnants on the hilltop. Another possibility for the creation of the gravel bar above the Emmanuel Church is the proposed Lake Mahanoy which may have been created by a blockage in the Mahanoy Creek water gap at Dornsife.

Wisconsin Ice Age rounded pebbles on a hill top east of Hunter Station.

The Dornsife Gap could have been blocked by ice or gravel during the ice age, creating Lake Mahanoy.

The Schwaben Creek joins the Mahanoy Creek near the Village of Red Cross. Tremendous gravel bars are deposited at this confluence. These deposits are identified on geologic maps as Pleistocene (Ice Age) in origin. If one views these deposits, it is possible to imagine how the valleys appeared 10,000 years ago.

Quaternary Age AlluviumIce age gravels deposited near Red Cross where the Schwaben Creek meets the Mahanoy Creek.

Dornsife Gap in Line Mountain: The Pocono Formation is cut by the mighty Mahanoy Creek which begins at Mahanoy City and meets the river at Herndon.

Pump Station Village, located between Dornsife and Red Cross, named for the pipe line pumps located here by John D. Rockefeller. The pipe line passes through the Dornsife Gap. The workmen lived here.

The Pump House built by J.D. Rockefeller. Steam was generated in a larger brick boiler house with a huge chimney. The boiler house is gone. Pipeline oil pumps were located in this building. During the Ice Age, the Mahanoy Creek flowed at a much higher elevation above the present stream level.

National Transit Oil Company, Latsha Station. A view from the Reading Railroad. For more information about this railroad and the pipeline see the book "Trevorton, Mahanoy and Susquehanna Railroad" identified in the bibliography.

Wehry Meats is built on an elevated terrace of Wisconsin Age gravel, higher than the low land which is frequently flooded. Mahantongo Creek, Pine Creek and Snow Creek join near by the butcher shop.

Another example of a more local Ice Age terrace can be found near Klingerstown where the Mahantongo, Pine, and Snow Creeks join near Wehry Meats. This butcher shop and farm has its buildings located on a terrace higher than the lowlands along the creek. Indian artifacts are known to be found on this elevated ground. Few arrow heads are found in the lowland. Further downstream on the Mahantongo Creek a gravel terrace can be found where Tim and Valerie Specht have a picnic pavilion. This building is on an elevated gravel terrace immediately adjacent to the creek. This terrain is only touched by water during severe episodes of flooding when brown sand is deposited in the pavilion.

The presence of sand in the Mahantongo Creek outwash continues today. During the last flood in Klingerstown the creek overflowed its banks and deposited several large truckloads of clean fine sands in the lowland fields east of Troutman Brothers. Some of this sand was removed and stockpiled for use in construction purposes.

Lowlands Between the Pine and Mahantongo Creek: Pleistocene Era gravel flood plain in Klingerstown, developed during the Wisconsin Ice Age.

Glacial Topography of Windblown Deposits

Ice age stream banks and flood plains along flowing waterways were bare of vegetation. Winds blowing across the rocky terraces picked up sands and deposited the sand in dunes. An example of a wind formed sand dune can be found in Paxton where the Mahantongo Creek joins the Susquehanna River. Louis Peltier states on page 49 of *Pleistocene Terraces of the Susquehanna River*, "On a hilltop ½ mile north of Mahantongo, at the confluence of the creek and the Susquehanna River, 139 feet above the river, there is a windblown deposit. The hilltop is gently rounded and on the north side near the top there is a shallow closed depression four to five deep and 150 feet in diameter. The hilltop is covered by a mantle 10 feet thick of very fine sand. The sand is in the upper 18 to 20 inches, buff colored, and below this, it is reddish brown in color, as though it had undergone Illionian weathering." Several years ago, Byron Rebuck built a new house on top of the old dune. Byron can tell the story of the sand excavated during the construction of his home. The sand was bottomless. His home is on the east side of Route 147, ½ mile north of the Mahantongo Creek. Byron's neighbors on the west side of Route 147 confirmed the depth of the sandy deposit also, having experienced that they can dig with hand tools at least 10 feet down without finding any bedrock.

Another similar deposit of windblown sand is along Wiconisco Creek near Millersburg. It lies on a bench 80 feet above the river. The reddish sand mass may be Illinoian and the brown or buff colored fine grained masses are almost certainly Wisconsin. The deposits are on the left bank of the Wiconisco Creek. (Louis Peltier, *Pleistocene Terraces of the Susquehanna River*, p. 49)

Bryon Rabuck's house is built on a wind blown sand mound high above the Susquehanna River at Paxton.

Glacial Topography of Hills and Mountains

We have discussed the effects of erosion by moving ice, water and wind during ice age times, changing the landscapes along streams and rivers. We can also see evidence of this severe climate on the hills and mountains. Louis Peltier describes on page 69, Boulder-Strewn Slopes. "Numerous surface features were produced by frost action. The tops of hills and mountains were rounded and covered with congelifractate. The slopes were smoothed and lengthened. Later surface erosion of the congeliturbate on the slopes led to the concentration of boulders on the surface, particularly in small stream channels. Block Fields and Boulder –Strewn Slopes were formed during both Wisconsin and Illionian times. It is sometimes possible to distinguish between the periods of strong frost action by the amount of weathering on the boulders." In our area study, the mountain ridges are all covered with a mantle of rubble derived from the sandstone bedrock by frost action. Fisher's Ridge and Hoofland Mountain are composed of Devonian Age Montebello sandstone. Fisher's Ridge near Klingerstown has on its south flank, a treeless rocky slope composed of boulders ranging in size from several feet to yards in diameter.

This rock glacier owes its origin to severe frost heaving of the bedrock during ice age climate conditions. A nearby fault mapped as Fisher Ridge Fault, may also contribute to this rocky slope. This site was utilized as a quarry during the depression era by the WPA doing local work projects. Many roads in the Klingerstown area have their berms fitted with these stones, cut by the workmen. The Tulpehocken Path Highway, west of Klingerstown, and the highway to the winery east of Klingerstown, have these stone berms. The wall along the Pine Creek in Klingerstown is constructed of cut stone quarried there also. The

Fisher Ridge rock glacier in Montebello Sandstone formed during the Ice Age.

Mahantongo Mountain and Line Mountain are composed of Mississippian Age Pocono formation. Coal or Thick Mountain, Big Lick Mountain, Mahanoy Mountain, Big Mountain all composed of Pennsylvanian Age Pottsville formation "are all covered by a layer of rubble derived from the massive resistant quartzite sandstone and conglomerate formations. These boulders tend to be concentrated at the surface and are the result of periglacial frost action being broken from the mountain tops and frost heaved down the slopes."

An example of frost heaved blocks can be seen in the Mahantongo Valley near Rough and Ready. The Leck Bush is a long narrow valley trending north to south in the center of the Mahantongo Valley. This area is locally known to be the site of a gold mining effort during the early 1900's. Immediately north of the mine location, just south of the intersection of Leck Bush Road and Pumpkin Center Road, extremely large boulders are concentrated along the Leck Bush Creek. The hillsides on both sides of the small stream are smooth without any evidence of rock on the surface.

However, more than a dozen large boulders lie within or adjacent to the stream channel. These large rocks are in size as big as a pick-up truck box or larger. The boulders can be seen on the west side of the Leck Bush road. Some of the boulders are actually on top of one another attesting to a tumbling movement from higher up on the slope. During the Ice Age this narrow valley must have been quite barren of vegetation with a bedrock cliff exposed to frost action. The boulders broke away from the exposed strata near the hilltop. Frost heaving and gravity moved the boulders down the slope. This movement occurred over thousands of years. At some point there must have been more rapid movement as two of the boulders of different shape rest on top of one another. They are perched adjacent to the stream channel. Thousands of years of weathering has smoothed the valley sides so that only the large boulders remain visible.

Leck Bush: Central Mahantongo Valley Smooth hillsides in this valley due to freeze-thaw cycles in the ice ages.

These boulders broken free from exposed bed rock in the ice ages.

Fallen boulders moved by freeze and thaw cycles from higher in elevation to the valley floor. Two rest fallen on top of one another.

Paleolithic-Recent Period

The first people arrived in our valley during the close of the Wisconsin Ice Age, 10,000 years ago. This would be in the last minute of the geologic time scale that we have been following. They were wandering hunters, traveling on the land near the edge of the retreating ice sheet. The environment in the valley was like that of northern Canada. Permafrost conditions existed for thousands of years. The only plant growth was grasses and scrubby trees. These native Americans are known to have camped in the Armstrong and Powells Valleys. One of their camps is known as the Shoop Site, which has been studied and documented as a hunter gathering place where meat cutting was done. A publication by Gary L.Fogleman describes this famous Paleo site. Jim Stutzman of Millersburg is a local expert on Indian lore. He suggests that these nomadic people followed herds of animals upon which they depended on for food. These Indians were hunting giant bison which stood 8 feet at the shoulder and had a horn span of 6 feet. Ice age animals roamed the valleys including bison, large bears, mammoths and migrating herds of caribou passed through the wind gaps in our area as they migrated north to south and south to north according to their seasons.

By 8000 BC, the climate had warmed and the Ice Age animals no longer existed. Pennsylvania was now covered with trees much the way it looked when our ancestors arrived. Archeologists look for rock shelters as evidence for the earliest settled places. A rock shelter dwelling has been located on the Swatara Creek near Echo Valley between Tremont and Pine Grove. The story of its recent discovery and the petroglyph stone face carving has been recorded by Francis J. Burke in "Schuylkill Living". (See bibliography for this magazine article)

Woolly Mammoths Mammuthus primigenius and Woolly Rhinoceros Coelodonta antiquitatis Cenozoic. Europe, Late Pleistocene time, 100,000 years ago. ID# CK30T Charles R. Knight paintings from the Photo Archives. All images shown in the gallery are © The Field Museum. Photograph by Ron Testa or John Weinstein.

Bringing Home Supper: Pine Creek Water Gap EncampmentThis could be the scene from the Rauch Gap near Valley View, or the Klingerstown Gap, near Erdman, 3000 years ago, looking north. Ink drawing by Jim Stutzman of Millersburg, from his publication, Indians of Dauphin and Schuylkill Counties.

"Spring Melt" by Jim Stutzman: This scene of an ice covered Susquehanna River could be the view from Millersburg. The Native Americans gathered along the river in winter seasons for easier access to fish and wild game.

Conclusion

In conclusion we have followed the geologic time scale from its beginning 4 billion years ago to the present. We have followed the local evidence in the layers of rock forming the Earth's surface in the Mahanoy, Mahantongo, Lykens, and Bear Valleys. We have traced the formation of the sedimentary rock layers, the process of uplifting and folding these layers to form mountains, and the subsequent erosion of these mountains to form the Ridge and Valley topography of our area. This process occurred over a period of 400 million years. It is difficult to express the tremendous length of time the geologic time scale describes for the development of our area. Mr. Jim Charles of Selinsgrove, an Isle of Que historian, uses a stick four feet long to represent the 400 million year history of the Susquehanna River Valley. Each foot represents 100 million years. The last ½ inch of the last foot, is when native Americans arrived. The last $1/16^{th}$ of an inch represents when Europeans arrived. Of the 48 inches on the measuring stick, 47 and $15/16^{th}$ inches are prehistoric. The last $1/16^{th}$ inch represents recorded history. European emigrants arrived by the mid to late 1700's in our area of study. One example is pictured below.

Peter and Eva Elizabeth (Meyer) Trautmann Homestead: Pioneer dwelling located south of Forest School House Community Center, between Hebe and Pillow, PA. Peter Trautmann was a resident of the Mahantongo Valley prior to 1775.Ink drawing by Charles "Chuck" Lappert of Akron, Ohio, now deceased descendent through Geo. Peter and Elizabeth (Geiss) Trautmann. George Peter was the son of Peter and Elizabeth (Meyer) Trautmann.

Closing

As it is recorded in the Bible, there were 7 days in the creation story. God created man in his image on the 6th day of creation giving man dominion over the Earth. And on the 7th day he rested. We find the geologic time scale to be compatible with the sequence of time in the Biblical story. The story of 6 days of creation by God is analogous to the 4 billion years of creation proved by geology.

Modern day astronauts who viewed our planet from space speak with awe as to its beauty. James Irwin described the Earth as a "beautiful, warm, living object, so fragile, so delicate..." Edgar Mitchell observed, "Suddenly from behind the rim of the moon, in long slow motion moments of immense majesty, there emerges a sparkling blue and white jewel...rising gradually like a small pearl in a thick sea of black mystery." "My view of the planet was a glimpse of divinity."

"The Home Planet" published by Addison-Wesley Publishing Company, 1988 photo number 45, conceived and edited by Kevin W. Kelly for the Association of Space Explorer.

Bibliography

Barnes, J.H., and Sevon, W.D., 2002, The Geological Story of Pennsylvania (3rd ed.): Pennsylvania Geological Survey, 4th ser., Educational Series 4, 44p.

Burke, F.J., 2001, (Spring), The Mystery of the Stone Face: Schuylkill's Indian Petroglyph, in Schuylkill Living Magazine.

Crowl, G.H., and Sevon, W.D., 1980, Glacial border deposits of Late Wisconsin Age in northeastern Pennsylvania: Pennsylvania Geological Survey, 4th ser., General Geology Report, 71, 68p.

Fogelman, G.L., 1986, Shoop, Pennsylvanian's Famous Paleo Site: A Popular Version.

Hoskins, D.M., 1969, Fossil collecting in Pennsylvania: Pennsylvania Geological Survey, 4th ser., General Geology Report 40, NN p.

Hoskins, D.M., 1999, Common fossils of Pennsylvania (2nd ed.): Pennsylvania Geological Survey, 4th ser., Educational Series 2, 18p.

Knorr, L.K., Troutman, S.E., et. al., 2013, There is Something About Rough and Ready, Sunbury Press, 50-A West Main Street, Mechanicsburg, PA 17055.

Leverett, F., 1934, Glacial Deposits Outside the Wisconsin Terminal Moraine in Pennsylvania, Topographic and Geologic Survey, Pennsylvania Survey, 4th Series, Bulletin G7.

Matthews, W.H., III, 1962, Fossils an Introduction to Prehistoric Life, Barnes and Noble Inc.

Peltier, L.C., 1949, Pleistocene Terraces of the Susquehanna River Pennsylvania, Pennsylvania Geological Survey, 4th Series, Bulletin, G23.

Sevon, W.D., Fleeger, G.M., and Shepps, V.C., 1999, Pennsylvania in the Ice Age (2nd ed.): Pennsylvania Geological Survey, 4th ser., Educational Series 6, 30p.

Shultz, C.H., 1999, The Geology of Pennsylvania: Pennsylvania Geological Survey, Pittsburgh Geological Society, Special Publication One.

Slatik, E. R., 1979, (June) Pennsylvania Mountains, Pennsylvania Game News publication.

Trexler, J.P. and Wood, Jr., G.H., 1968, Geologic Map of the Klingerstown Quadrangle, Northumberland, Schuylkill, and Dauphin Counties, Pennsylvania, U.S. Geological Survey Geologic Quadrangle Map GQ-700, 1 oversize sheet. Available at http://ngmdb.usgs.gov/Prodesc/proddesc 2008.htm .

Troutman, S.E., et.al., 2007, Klingerstown Bi-Centennial Album, 1807-2007, published by the Anniversary Committee, Heule Gordon Inc., 1415 Plainfield Ave. NE, Grand Rapids, MI 49505.

Troutman, S.E., The Trevorton, Mahanoy and Susquehanna Railroad, Sunbury Press, Inc., 50 West Main Street, Mechanicsburg, PA 17055.

Troutman, S.E., 2011, Tulpehocken Trail Traces, Sunbury Press, Inc., 50 West Main Street, Mechanicburg, PA 17055.

Willard, B., 1965, Pennsylvania Geology Summarized, Educational Series, No. 4, Topographic and Geological Survey.

Wood, G.H., Jr.; Trexler, J. Peter; Kehn, Thomas M., 1969, Geology of the west – central part of the southern anthracite field and adjoining areas, Pennsylvania. U.S. Geological Survey Professional Paper 602, 150p. http://pubs.er.usgs.gov/publication/pp602

Appendix

Geologic map of the central portion of the Mahantongo Valley. USGS GQ-700 by J. Peter Trexler and Gordon H. Wood, Jr., 1968.

Geologic map of the central portion of the Hegins-Lykens Valley and Bear Valley. USGS GQ-701 by J. Peter Trexler and Gordon H. Wood, Jr., 1968.

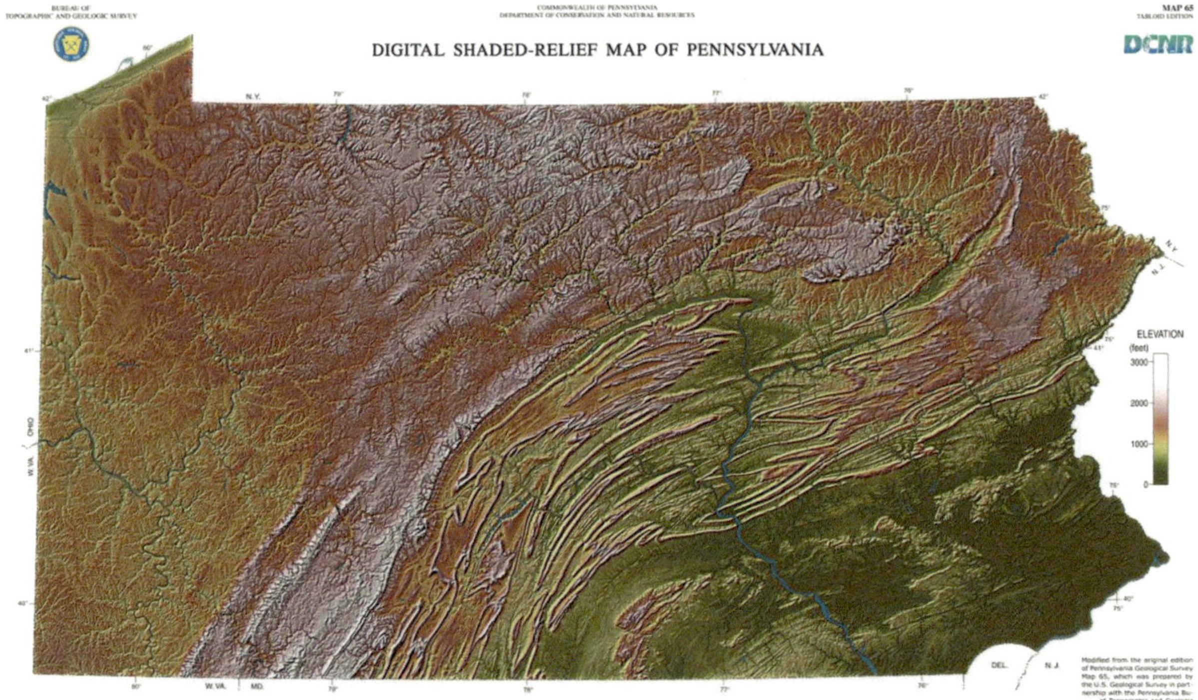

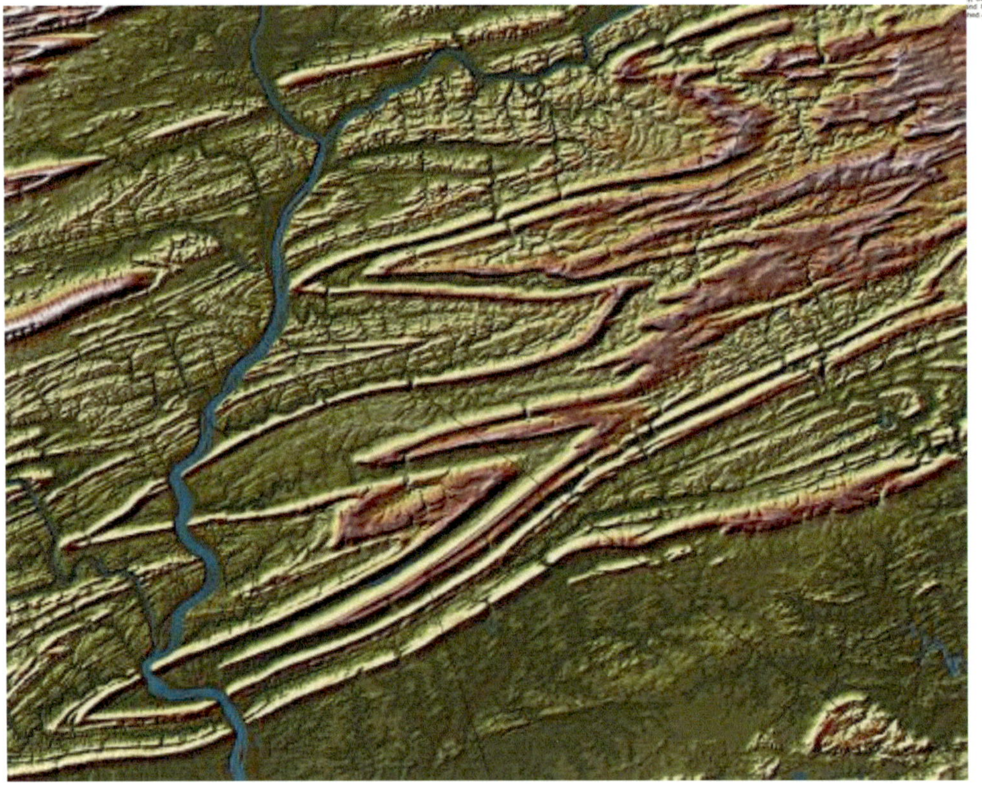

Enlargement showing the area of our study including the Mahanoy Valley, the Mahantongo Valley, Lykens-Hegins Valley, Bear Valley, and the Powells-Armstrong Valley.

The Spread Eagle Tree in Klingerstown, PA recreates an Indian carving on a sycamore tree. The tree was located on the Shamokin-Tulpehocken Path. Moravian missionary Zinzendorf's map of 1742 depicts the tree location on the trail. This Indian path utilized three water gaps and out flanked several ridges. It was used as a direct route between the Indian Village of Shamokin at Sunbury and the Tulpehocken Valley near Reading.

The old Klingerstown Mill. Alexander Klinger's son, John, built this log mill before 1800. According to archaeologist Fred Kinsey III, it was the oldest mill in the Mahantongo Valley and perhaps one of the oldest in Schuylkill County. The red brick grist miller's residence was built in 1833. The mill was able to utilize water from the Mahantongo and Pine Creeks. Dams were built on both creeks with mill races to each stream. The Mahantongo Valley and Lykens Valley water sheds provided an ample supply of water power.

Zion (Klinger's) Church at the Pine Creek Water Gap in Mahantongo Mountain. The creek's northerly flow may have had its direction established in the Alleghanian Orogeny when mountains were uplifted southeast of Pennsylvania.

Pioneer Philip Klinger first lived in a dugout hut built into this hillside of Mississippian Age Mauch Chunk Formation of terrestrial red shale.

Luxemburg Bridge, Erdman, Pa. being rebuilt over the Pine Creek in 2013, near Peter Klinger's mill site. This stone arch bridge was severely damaged by flood waters. Peter Klinger's grist mill race carried water under the bridge through a separate arch to provide power to the mill. Flowing water can by a source of energy but it is often a powerful force of erosion as mentioned often in this book.

Pillow Gap looking north from the crossroads west of Gratz. The Deep Creek flows in a northerly direction through the gap in Mahantongo Mountain. The creek's northerly flow may have its origin in the drainage pattern established during the Alleghanian Orogeny, when mountain building occurred southeast of central Pennsylvania. These uplifted mountains were the source of stream water which flowed in a northwesterly direction.

The Author

Author Steve E. Troutman holding a pillow structure, August, 2014. Pillows are rounded masses of solid rock formed from the tumbling movement of mud generated by underwater mud slides. These large pillow structures were found east of Pitman, Pa., in the Catskill Formation of Devonian Age.

Made in the USA
Las Vegas, NV
11 November 2026

57876074R00045